THE EXTENDED REALITY BLUEPRINT

ANNIE EATON

THE EXTENDED REALITY BLUEPRINT

DEMYSTIFYING THE
AR/VR
PRODUCTION PROCESS

WILEY

For general information on our other products and services or for technical support, please contact
our Customer Care Department within the United States at (800) 762-2974, outside the United
States at (317) 572-3993 or fax (317) 572-4002.

Wiley also publishes its books in a variety of electronic formats. Some content that appears in print
may not be available in electronic formats. For more information about Wiley products, visit our
web site at www.wiley.com.

Library of Congress Cataloging-in-Publication Data is Available:

ISBN 9781394207688 (Cloth)
ISBN 9781394207695 (ePub)
ISBN 9781394207701 (ePDF)

Cover Design: Wiley
Cover Image: © Shabira line icon / Adobe Stock
Author Photo: Haigwood Studios
SKY10067131_021424

To my Futurus team past, present, and future,
who teach me something new each day.

Contents

Acknowledgments

WRITING THIS BOOK has made me reflect on the past decade and what I've learned. The entrepreneurial journey is difficult, lonely, and trying, filled with the highest of highs and lowest of lows. My husband, Jake Lance, has been with me since before day one, advocating for me and supporting me like no one else could. I love you, and you mean the world to me. I'm privileged to share my life with you. Starting a business would have only been a dream if it weren't for the partnership of Peter Stolmeier, who taught me so much in the world of extended reality. Thank you for your friendship and for sharing this transformative technology with me. And Futurus could not have become the incredible brand it is today without the dedication of Amy Stout, who has become my most trusted advisor and closest friend over the years.

I want to thank my family for their unconditional support and love. I am truly blessed to have Danice and Rick Eaton as my parents. They never doubted what I could accomplish, even when it sounded crazy to the rest of the world. And to my siblings, Richard Eaton and Holly Arencibia, thank you for putting up with me all these years. You helped shape me into a strong person, and I'm inspired by both of you every day. Thank you also to Dirk Rountree, Paul Welch, Chan Grant, David Macias, Elijah Claude, Victoria Savanh, Adaobi Obi Tulton, and Venkatasubramanian Chellian for helping me make it to the finish line with my first book. To the rest of my family, friends, and peers, your support as I traverse through the entrepreneurial journey has been incredibly meaningful, and I appreciate each and every one of you. I would not have gathered even a fraction of this knowledge if it weren't for the people in my life that have been along for this ride with me.

About the Author

Annie Eaton is an immersive content producer, specializing in engaging and interactive virtual reality and augmented reality experiences. She is the founder and CEO of extended reality-focused company Futurus, which produces training and product visualization applications and provides technology consulting for various organizations and nonprofits. In addition to her current role as CEO of Futurus, she also serves as executive producer of Amebous Labs, a virtual reality–focused game studio, publishers of Loam. She stays involved in her local Atlanta technology community, managing XR Atlanta, an organization that supports thousands of extended reality enthusiasts and professionals across the city and beyond. Annie is also involved with Women in XR Atlanta, Women in Technology (WIT), and The Academy of International Extended Reality (AIXR) and serves on multiple industry advisory boards.

Prior to her work in extended reality, she earned her degree in international affairs and modern languages from The Georgia

Institute of Technology. She initially entered a career in marketing and communications, where she was first introduced to virtual reality technology. After being exposed to early virtual reality hardware and experiences, she made the leap to enter the field of immersive technology and forged her own path through entrepreneurship. Annie has found a way to blend her creative, communications, and technology experience to build an innovative team that tackles the latest challenges in extended reality production.

She has been recognized by various organizations for her work in tech, including Women in Technology's Woman of the Year Award, Women in IT's Rising Star Award, and Technology Association of Georgia's Young Professionals' Technologist of the Year Award. Her projects have won recognition from Fast Company in their Innovation by Design award series. Annie has vast public speaking experience, having been invited as a speaker for conferences, podcasts, educational institutions, and corporate events to share her love and knowledge of immersive technology. She resides in Atlanta, Georgia, with her family and dogs. *The Extended Reality Blueprint* is her first book.

Introduction

MY EXTENDED REALITY career was formed at the intersection of hard work and happenstance. As with many of my colleagues, my interest in extended reality, or XR, came out of nowhere. It happens one random day when someone is introduced to the tech for the first time, and the next day, they're obsessed. This is my story: one day in 2013, I was introduced to the Oculus DK1 virtual reality headset (at the moment, the most successful Kickstarter of all time) by my coworker Peter, and I was instantly hooked. There weren't many resources to learn about extended reality for people like me—non-developers—other than academic journals and Reddit, polar opposites. So, with the encouragement of that same coworker, I began to read everything I could get my hands on.

This went on for months. I even got the opportunity to interview at Oculus, which took course over several months in 2014, right on the heels of their acquisition by Facebook. I can honestly say (now) that I am grateful that didn't happen. That was probably the best "no" I've ever gotten in my life because it made me want to prove them wrong. There was a place for me in the industry, and I was going to prove it. And nearly a decade later, I am confident to say I did.

Before any inkling of a company was formed, I needed to network in my local area of Atlanta, Georgia. That same coworker and I decided to start a meetup group, and it had an unexpected organic growth that proved to me that my city was ready to accept virtual reality technology. The "strategy" behind the first event wasn't much at all. Peter and I created a group called Atlanta Virtual Reality Meetup on meetup.com and posted our first event with the title "Try out some exciting demos and talk to others about virtual reality." That's about the strangest event title I've ever heard, but it worked. In addition to sharing with friends and posting on Reddit, the venue that allowed us to set up shop sent out the event to their mailing list. We didn't realize what a powerful mailing list a gaming bar would have, but nearly 200 people showed up (triple the amount that RSVP'd), and there was barely any room to walk around—it was overcrowded. People loved it, and they wanted more.

It was in that moment I knew this opportunity was much larger than I realized. Peter and I had to figure out what to do after the success of this gathering. People were asking questions, and we needed to give them answers. We attended other local events to demonstrate the technology, and my name started to be associated with virtual reality in Atlanta. Strangers were contacting me to ask how their organizations could leverage virtual reality. I'd become the posterchild for the virtual reality community in Atlanta almost overnight. I had no intentions of starting a business when this whole journey started, but I was also not going to ignore the slew of inquiries coming my way. I did want to work in the industry after all, and maybe this was my chance. So, after a meeting with Peter and a woman who had reached out to me asking if we could provide some guidance, he and I stayed after in the coffee shop parking lot. We talked about this crazy idea of starting a business to help people find their virtual reality use cases. Prior to the release of the Oculus DK1, headsets were so expensive they were limited to industries such as military and medical. The emergence of cheaper hardware meant access for all, and we wanted to help everyone find a way to use it.

I bring all of this up to say that I didn't have a computer engineering background or years of creative agency experience prior to getting into my field. This industry is so ripe with opportunities that with an idea and the tools to bring it to life, anyone can become a creator. Yes, it was an immense amount of work to get where I am today, but it came from an inspiring moment and a vision of future possibilities.

There are so many aspects of the work that I do that I had to just figure out. The budding network of extended reality enthusiasts was a close-knit community when I started. Heck, it wasn't even called XR until there were so many digital realities that we needed a consolidated term. I asked questions when I could, but most outcomes were a result of intense Google searches and trial and error. My goal with this book is to provide a jumping-off point, a resource, an account of lessons learned, so that you can start with a leg up. This book consists of nearly a decade of findings and best practices summed up in one place, by someone who has tried all of it. And why share all this information when I'm still using it to grow my company? The technology is only as good as its adoption rate. The more people that produce effective, comfortable, and immersive content, the faster the tech will grow.

There isn't a place for any of us without an audience, and there is no audience without good content. The bottleneck isn't in the hardware—it's in the software and experiences that make the hardware great.

Definitions are important, but they are also ever changing. There are three that I will use the most throughout this book: virtual reality, augmented reality, and extended reality. This is how I define them as they pertain to this book and my work:

Virtual Reality (VR)

Technology leveraged to immerse oneself in a computer-generated 3D world through use of a head-mounted display, completely occluding the view of the physical world.

Augmented Reality (AR)

A digital overlay of 2D or 3D content onto the physical world, typically experienced with a smartphone, tablet, or wearable device.

Extended Reality (XR)

A universal term inclusive of immersive technologies such as virtual reality, augmented reality, mixed reality, and spatial computing.

This book will go through my process from the inception of an idea through its deployment. But before we can go through the process, there are some barriers that need to come down with how we think of digital content to begin with. We've been taught most of our lives to think on paper or on a flat screen, and those mental habits need to be shattered. The medium of extended reality gives all of society a new way to share their knowledge, creativity, and ideas. It is extremely difficult to come up with anything original, and now we have an opportunity to do so. Growing up, my uncle would bring up his "Society of Original Thought" or SOOT as he calls it. He's talked to me about it since I was very young, and the idea is to challenge yourself to come up with original ideas. It seems nearly impossible as there have been limited opportunities for original ideas and applications due to the inundation of products, services, and content for anything and everything. But with extended reality technology, we get to drop the norms and guidelines that the physical world constrains us to. We need familiarity somewhat for adoption but do not need to limit ourselves to what we already know and what is possible in the physical world.

Once these mental barriers have been removed, we can go forward with the production process. We must start at the end deployment to choose which technology type within the realm of XR we want to leverage. Making that decision at the start of a project has significant benefits when it comes to development tasks. This book will walk through exercises to help you choose the right technology for the use case, which will save a great deal of production time if determined early.

Once that deployment method is determined and a use case is clearly defined, you're off to the races. We find that over half of the process is design, documentation, asset gathering, and planning. It seems excessive, but the clearer a vision the development and art team have, the faster they can produce something incredible. If documentation is disorganized, inconsistent, or missing pieces, there is so much more back-and-forth that takes place and wastes time. Design is the most important part of any project, and it's also viewed as the least important by many (most often by my clients). Stressing that importance and baking it into every schedule, meeting, and discussion will build good habits from the start.

By laying a strong foundation, incredible creativity will shine through. When I first started my business, I thought processes were creativity-crushing vehicles. But it was only when I started putting processes in place that I didn't have to think as hard about the mundane repetitive aspects of work and was able to think more freely. Extended reality technology has plenty to offer, but that can be overwhelming. Leveraging processes to help guide openness and opportunity will get you to a workable idea much faster. The remainder of the book focuses on getting to that end goal of a successful deployment. From user testing to iteration, finalizing interactions to quality assurance, the many components that go into delivering an exceptional experience seem endless. This book will distill that down into manageable chunks to make the seemingly impossible possible.

I have learned the hard way how difficult it can be to produce extended reality work, and I wanted to create a blueprint to help others learn from the challenges I've been through. The demand for this type of immersive content is only going to grow, so we need more producers, developers, artists, and designers. I truly believe that anyone from any background can find a place in the field of extended reality. But to make that transition, we need XR champions—enthusiasts of the technology who believe in powerful possibilities and opportunities. When you hear about immersive training simulations that are saving lives, the ability to visualize something magnificent that doesn't yet exist, or narrative experiences that allow people to walk in each other's shoes, it makes you believe in the power of extended reality. Will you join me and change the way we see the world?

1

Perspective

LET'S START WITH an exercise. Hold your index finger out in front of you. Now draw a cube in the air. Think about how you drew that cube: Did you draw a flat cube with a square and diagonal lines to simulate a three-dimensional drawing on a two-dimensional plane? Or did you outstretch your arm to give your cube depth, extending into the third dimension? Chances are if you are an adult, you drew a flat cube. When I try this exercise with young children, they don't have the limitations of thinking about 3D objects as a 2D representation. They think about things how they are in the physical world. We have spent our entire lives viewing content on flat paper, a computer screen, a phone, and it's very difficult to reverse that mentality. But my hope for you is that you can train yourself to think like a child. To think in a more physical way. Because to create virtual worlds, we must drop much of what we've been taught all our lives. We need to remove the limitations that decades of schooling and society have engrained in our minds. It can be done, but it takes work to break those mental habits.

When we work in extended reality, many of the constraints that exist for other media are lifted. The laws of physics no longer exist (unless you want them to). Some of the best features of development engines such as Unity and Unreal are that they simulate physics—gravity, collision, motion, etc. However, these programs also allow you to turn physics off or adapt the experience in a way that is out of this world. Imagine a virtual reality simulation that has the gravity of the moon. Objects react in this experience as if they were on the moon when you throw them, bounce them, drop them. The only component that doesn't react in this way is your actual body (unless you're using some sort of harness to help simulate that too). While many extended reality experiences need to retain some of the earthly laws of physics, it's fun to think outside the realm of what's possible in the physical world and leverage such a powerful medium to achieve just that.

Physical versus Digital

When designing for extended reality, think in 3D. Drop the limitations of your 2D computer screen, phone, and tablet. Even if technically we will still be experiencing the content through flat screens, the way we perceive it is with depth and dimension because of the way it's being anchored or positioned in a space. So what does this mean, to have the freedom of depth and dimension? It means that we can visualize data and content in a way that is immersive. We can interact with content like we interact with objects in the physical world. But in XR, they don't even have to be objects that are bound to the physical world. We can create functional, technological magic.

You'll notice throughout this book that I do not refer to things as "real" and "fake" but rather "physical" and "digital." This is because the line between real and fake is blurring, and definitions are changing. If someone joins a social space and attends an event on their virtual reality headset, that is still a real experience to them. They made connections, conversed with people across the world, and created a memory—nothing fake about that. Differentiating between extended reality content, or digital content, and the physical world that grounds us isn't made up of real and fake. It's all real in some way, shape, or form. It's different from what we've always known, but that doesn't make it less real.

Sometimes it's a good thing to be able to replicate the physical world. It's not all about creating a digital fantasy where there are no boundaries. Many of the best use cases come from replicating parts of the physical world and their processes. Think of educational content and practice simulations for surgeons. By recreating what's in the physical world these doctors can practice indefinitely and with unlimited resources. There isn't any waste, and there is no liability or potential loss of life with a virtual practice simulation. Simply put, you can practice until you're fully prepared. Another great reason to replicate the physical world is to grant access to culture, travel, and art, for people who don't have the ability or means to access it in person. While in many cases it's hard to compete with the real thing, extended reality can give exposure that can lead to greater understanding around the world, and it's no longer limited to mobility, ability to travel, or socioeconomic status.

Accessibility

Accessibility in the realm of extended reality still has a long way to go, and there is much to be desired, but there are some aspects of accessibility that are made possible by this technology. Leveraging positional data, or where someone is located in a space, and then being able to give detailed audio cues and information based on that positional or spatial data is a powerful way to assist those who are visually impaired. The ability for a program to intake real-time audio and generate voice-to-text displayed directly in the line of sight of an individual via their glasses is an effective way to assist someone who is hard of hearing. Taking that same technology one step further and generating real-time translations could help break global barriers in communication. The ability to lower the interactive components of an environment with the click of a button can help someone who can only experience the world in a seated position. The possibilities for accessibility in these virtual spaces and with augmented tools are going to open a wide array of opportunities, efficiencies, and quality-of-life assistance as never seen before.

Possibilities

So what is possible in XR? When I say the possibilities are endless, that may in fact be true based on the human imagination. However, while there are some capabilities that stand out as winning features, there are limitations as exist with all technology. Starting with digital environments and assets, it's possible to make a 3D space that looks like anything you can think up. By leveraging 3D modeling tools such as Blender or Maya, the creative limitations are bound to your technical modeling skills. Thinking of creating a futuristic otherworldly inhabitation? That can be done! When planning these spaces, ponder about everything you could ever want in an environment. How do you eat? How do you use transportation? What do the trash cans look like? Whether it's a fantasy world or very much rooted in the physical world, this level of detail is crucial. The same goes for single objects that may be experienced one at a time or overlaid onto the physical world. And while creative limitations are boundless, unfortunately, technological limitations exist. Environments and art assets will need to be optimized for the device on which the content is being deployed, so keep that in mind along the way, and consider where you can make compromises if necessary.

Interactivity

Interactions are another imperative factor of any extended reality experience. It's important to think about how to implement interactions in these worlds that you've created or with these digital assets you've produced. Interactions are essential and allow designers to bring what may be impossible in the physical world into a functional virtual instance. Interactions and the complexity of these interactions have a vast range, but the most common is "grab." The act of picking up an item and placing it,

moving it, throwing it, stacking it, dropping it is not only a nov-
elty but a requirement in most extended reality experiences.
While it seems the simplest, if "grab" doesn't feel right, it can
ruin an experience. "Grab mechanics," as they're called, are common across simulations in their implementation, and extended
reality users have grown to expect a level of consistency across
experiences, which is good for the industry and its progress. Interactions can be made using a controller or gesture/hand tracking
depending on the hardware's capabilities.

Moving beyond "grab," most other interactions will involve
some sort of physical intersection of the "hand" or controller
and the digital object. Think of outstretching your index finger
and pushing a physical button with it. That same interaction
can be translated into a digital experience using that exact
implementation. And a third type of interaction would be user
interface, or UI, interaction. This is less immersive, but it can
feel more familiar because this is typically used to access menu
screens or dialogue boxes, similar to 2D digital content. This is
commonly done by using a point-and-click method, like what
you'd experience with a computer mouse or even your finger on
a phone screen. Interactions are a crucial consideration in any
extended reality experience, so take advantage of the possibilities and give thought to implementation.

Movement and locomotion add to the opportunities that extended
reality brings to its audience. Moving the user in your experience
can be tricky, and detrimental if not considered with care. If you
move the user without their consent, it can be nausea inducing,
which is an immediate turnoff to the technology in general. On
the positive side, in extended reality you're able to move in ways
you couldn't in the physical world. Teleportation is one of
the most common forms of locomotion. I know you're thinking,

"But isn't that some 'Beam me up, Scotty!' science fiction?" Well, in virtual reality, blipping from one area to another is as easy as the click of a button. Since walking can be difficult due to limited physical space, teleportation has been adapted to not only take you from one environment to the next but similarly to hop quickly around a single space. It's my favorite means of movement in an extended reality experience and can be quite fun once you get the hang of it. For the iron-stomached audience playing fast-paced games, a common method referred to as "smooth locomotion" is often used. When you think of smooth locomotion, think of maneuvering with a joystick in any direction that you move it. While it does provide the user with more flexibility and precise movements, this can also induce motion sickness since the visuals of the experience move quickly in the peripheral while the user remains stationary.

Branching out into more creative methods of movement, some experiences use motions that mimic swimming, swinging, and even gorilla arms to trigger locomotion. Notice one thing in common—all three of these examples are using arm movements to propel the user forward. This is due to the fact that legs and feet aren't tracked yet. In virtual reality specifically, we're limited to a headset and two controllers. There are third-party haptic devices that can be added to the setup for additional tracking, but designing for the hardware that comes in the box is important to reach the widest audience. Therefore, the developers who get creative with locomotion commonly use arm movements to locomote.

Human Connection

Digital social connection is everywhere we look (we can't get away from it, even if we try). I will proudly say that the most digitally connected I've felt to someone has been through extended reality. While social or multiplayer features aren't a fit for every experience, seeing a person's avatar in 3D, face-to-face, and up close is an immersive element that can only happen with extended reality. It's different than Bitmojis and Snapchat filters you'd see on your phone or computer screen. You can experience the mannerisms of the person in front of you just by the micro-movements of their head, arms, and hands. In 2016 a social VR platform called AltspaceVR was doing a road tour and stopped in my city. I helped set up an event with them to draw out local virtual reality enthusiasts to give them feedback as they were continuing to evolve their product, which eventually was sold to Microsoft (and later, unfortunately, sunsetted). My mother stopped by the event to check it out. During the virtual reality demo, she was talking to another person, carrying on and having a fantastic conversation. It was only when we pulled her out and explained the platform that she realized she had been conversing, laughing, and engaging with people from all over the world. She got a little upset after she learned this, as she realized she would never see or talk to her "new friends" again. She was able to make an emotional connection over virtual reality with strangers over a five-minute tech demo. I play games monthly with my team (several of whom live in other states) in virtual reality, and I am appreciative of how close I feel to them even though we aren't together geographically.

Social experiences in extended reality are powerful and have many positives, but just like any social platform, moderation and personal safety controls are necessary. Harassment still happens, but common moderation features in public spaces such as muting or blocking other players help. An XR-specific safety tool is the "personal bubble" setting, which many platforms are starting to adopt. When this setting is on, anyone who comes closer than an arm's length will automatically be invisible to the player.

I especially appreciate this in public social events, and I sometimes wish this existed in the physical world!

Avatars

One of the most fun components of these platforms is visual self-expression, commonly in the form of avatars. An avatar is your personal character that you can design and customize, which will serve as your "body" in an extended reality experience. Avatars don't even have to represent a human form, with options such as robots and anthropomorphic characters. Most platforms have their own avatar style and therefore their own customization menu, but the hope in the future is that an avatar would be cross-compatible among all experiences. There are already companies working on this, such as Ready Player Me. A completely customized avatar isn't always necessary, however. My team designed an industrial-focused classroom experience, and the "character" is represented by a hard hat and safety gloves, with each person assigned a different color. The trainee's name is above their head to distinguish who is who, but this method reduces distractions in the classroom while still giving the trainees the feeling they are in the digital space with other people.

Presence

Regardless of what you design and how you implement it, you'll want to consider presence, which is a term that has become synonymous with a good extended reality experience. Presence is the feeling you get when you're in the experience—the feeling of really being there. It's a difficult word to define, but for example, if you're in a virtual reality experience and feel like you're truly immersed in the virtual world (perhaps even lose track of your position in the physical world), that means the experience has presence. I would consider presence a subjective metric for evaluating an experience. It's a feeling that you need to design for, which we'll dive into throughout the design chapter.

Determining the Medium

Most of the possibilities discussed have an option for both augmented and virtual reality implementation, depending on how you're going to release the content. So why would someone choose augmented reality versus virtual reality? Augmented reality is the best fit when you want to enhance the physical world. In most cases, you're not creating something completely new, you're adding information or content on top of what already exists physically. If you want to view products in your own space prior to purchasing them, see what steps you need to take to make a physical machine work, or experience an advertisement come to life, augmented reality is the best path.

Virtual reality is better suited to immerse the user in a completely new or digitally replicated environment. When using virtual reality, the physical world is visually occluded, so use that to your advantage. If you want to show someone a building that hasn't yet been constructed, train an employee on a piece of dangerous machinery, or take a trip to Mars, virtual reality would be the best medium to consider. Determining the use case and identifying the hardware for your experience will help you make the decision of which path is the best fit. Whichever path you choose will lead to incredible, immersive content, so enjoy your journey into extended reality production and remember to think like a child.

2

Hardware

WHILE HARDWARE IS one of the most important components to any experience, it's also one of the most complex choices to make early in the process. Making the choice of what hardware to support guides the entire product journey and is more challenging to adjust later in the process if the decision is made in haste. Are you creating for one device? Do you need cross compatibility? Will your app have a complementary companion experience? In the future, there is going to be more crossover between virtual reality and augmented reality, hopefully in one shared device, which is sometimes confusingly referred to as mixed reality. We'll define them separately for now.

Augmented Reality Hardware

Augmented reality can be accessible from personal mobile devices, such as phones and tablets, as well as wearables in the form of glasses. You can still see the world around you but with a digital overlay of content onto your physical space. When accessing augmented reality from a phone or tablet, it's like looking through a window into a digital world. You can hold your phone up, scan the environment around you, and then content can be mapped to planes (such as the ground or a table), an image target (such as a menu or advertisement), or even your own face or body. A combination of camera mapping and accelerometers in mobile devices help your phone know its position so that as you move around, the content stays mapped to the physical world.

While phones and tablets are currently the most common way to access augmented reality, wearables are growing more popular in the enterprise space and are advancing in the consumer market as well. The two most popular headsets at the time of writing—Microsoft HoloLens and Magic Leap—have sophisticated spatial mapping that maps more than just flat surfaces. They can map entire facilities, advanced shapes, and even people. Leveraging similar spatial mapping technology, consumer devices are starting to emerge, such as the XREAL line of wearables. And newer enterprise devices from Apple and Lenovo promise spatial mapping as well. For less immersive heads-up display use, Vuzix and RealWear are top contenders. With each new iteration of these devices come advancements in quality, field of view, interactivity, and comfort.

Virtual Reality Hardware

Virtual reality (VR) content is experienced through a head-mounted display, or HMD, which is worn on the face and affixed with a head strap. Virtual reality devices completely occlude the user's view of the outside world unless they're in "passthrough" mode. Passthrough allows the user to see a feed of the outside world through cameras (sometimes black and white and now more commonly in color). This started as a safety feature but has evolved into a functional way to blend augmented and virtual reality content into one device.

Virtual reality hardware can either run off a high-powered computer or remain completely portable. When virtual reality is running on a computer, it's often called "PCVR" or "tethered" due to the cable that connects the headset to the PC. These tethered systems can provide better quality experiences due to the graphics and processing power associated with a PC. A completely portable virtual reality system, often referred to as stand-alone, is appealing to individuals who want an easy setup and lower price point. The mass market seems to prefer stand-alone virtual reality systems as seen with the popularity of Meta Quest devices. The enterprise market is picking up on stand-alone systems, such as PICO, HTC, and Lenovo devices, which are easier to scale within an organization. By not requiring a gaming-quality computer, the barrier to entry is lowered.

Both AR and VR solutions have hardware options that could require controllers, or hand tracking technology will suffice. And then there is the optional addition of haptic devices, which allow the user to feel vibrations, and peripherals that support the ability to walk infinitely, or even smell their surroundings. Many important factors go into making a choice of which hardware to work with, but I typically start with defining three factors: audience, deployment method, and content. Once these are identified, it's much easier to make a hardware choice that will stand up for the duration of the project.

Audience

Depending on who your audience is, there are choices that rise above the others. If you're producing an experience that is consumer focused, identifying the most popular devices in people's homes is a great place to start. Releasing an experience on a device or system with a limited user base might not get you the eyes you're hoping for. For a consumer release, the widest augmented reality reach will be on personal mobile devices. While wearables are available to consumers, choices are still limited, and adoption is low due to the lack of content and price point of the devices. For a virtual reality consumer release, the Meta Quest line of devices is overwhelmingly the fan favorite, making up most of the consumer purchasing. Depending on the content, a PlayStation VR or PCVR release could also be a fit.

If the experience you're producing is business or educational focused, key in on devices that are manufactured for enterprise use.

There are headsets and wearables that are specifically made for enterprise deployment, such as the PICO 4E, Vive Focus 3, Lenovo ThinkReality VRX, Microsoft HoloLens, and Magic Leap. Hardware manufacturers are mindful of security concerns, scalability, and company-required controls. While these headsets may not be as popular in the mass market and are often more expensive, they are a fantastic choice for companies and organizations who want to release internal content and control the hardware themselves.

Deployment Method

Deployment is an important process to outline at the start of a project. This means how your content is going to get to the end user. The most common would be some sort of storefront, comparable to the App Store or Google Play on your phone. Meta and PlayStation have their own stores, which only support their own devices. Meta's is broken into two tiers, allowing experimental or limited versions of experiences to be released before launching to their full store. Steam is one of the most popular distribution platforms for all gaming content, and SteamVR supports most PCVR devices, such as the Valve Index, HTC Vive, and Varjo XR-3.

When enterprise or educational content comes into play and you're leveraging a virtual reality stand-alone headset or augmented reality wearable, a mobile device management system is almost always the best fit. There are several on the market to choose from such as ArborXR or ManageXR. These platforms allow for total control of deployed content and often provide security options to lock certain device features. These platforms typically have a web-based backend portal, where a systems administrator can group devices, deploy specific experiences to groups wirelessly, and monitor the devices remotely. For testing purposes or for a small batch of devices, "sideloading" is another popular option. However, to sideload content onto a device, you must plug the device into a computer for file transfer. The simplest way to sideload is by using a program such as SideQuest, which supports this feature and allows for easy file management while the device is plugged in.

Content

In the trifecta of hardware factors, content is the most limiting. Mobile devices, whether virtual or augmented reality, can only handle so much processing, which will limit the experience you are creating. In the case that you need to visualize (close to) photorealistic renderings, a Meta Quest just isn't going to cut it. This use case would require the processing power of a high-end gaming PC paired with a headset with amazing optics, such as the Varjo XR-3, to get the job done. If you're making a training simulation that consists of operating a piece of machinery in an empty warehouse, your options open widely and mobile devices such as the PICO 4E or Lenovo ThinkReality VRX may be a great choice.

On the augmented reality side of the spectrum, is it important that this content is hands-free? If so, a wearable device is a better option. Are you creating product visualization tools or social filters? Mobile phones and tablets are going to be a perfect fit. Some of the more industrial content integrates voice commands, so if that would be helpful, a heads-up display such as Microsoft HoloLens, RealWear, or Vuzix device would be best. Regardless of your choice, ensure that you're thinking about the content you're creating and what device is going to be the best for the user experience.

Cleanliness

Nothing irks me more than when I'm at an event and someone hasn't cleaned a publicly used virtual reality headset. It's the simplest thing that makes or breaks an experience, so I'm going to get on my soapbox and stand my ground. There are a few steps in the process of cleaning a headset, some required and some optional. As soon as I receive a device in hand, I confirm what the facial interface material consists of. Many arrive with a foamy absorbent material, which is okay for personal use but not a good idea for a shared device. Some devices come with a rubber or wipeable cover, and others require an aftermarket addition from a product line such as VRCover. Regardless of which option you choose, it's incredibly important to have a wipeable, cleanable facial interface if the device is going to be shared (or even if it isn't).

Once that has been established, you'll always want to have disinfectant wipes available. I personally prefer the single packet medical-grade alcohol wipes. These are easy to pack when you're taking the devices on the road, and I keep a stash at home, the office, and in all my cases. The purpose of using these is to wipe makeup, oil, or other residue from the facial interface. It's also a good idea to wipe down the controllers and the rest of the headset, but some prefer full-size wipes for that job. Whatever you do, do not use these types of wipes to clean the lenses or cameras. Use a microfiber cloth so as not to damage the delicate components.

Those steps are the bare minimum, but depending on your environment, setup, and needs, an ultraviolet light is an excellent final step to ensure optimal cleanliness. These devices can be purchased from companies such as CleanBox, and only take about a minute to use. While there are a variety of models and capacities, the principle is the same. Place the device in the box, press a button, the UV light works its decontamination magic, and you're ready to pass the headset off to the next person.

Comfort

After cleanliness comes comfort. Like the aftermarket options that can be purchased to keep devices sanitary, there are also adaptations that can be made for comfort. While not all the head straps can be changed out, some brands do provide flexibility. This can be helpful if you have a larger head or a specific hairstyle that is not conducive to a strap running across the top of your head. Manufacturers are now starting to design with an optional middle strap, or even no middle strap at all. I always prefer to change out my strap with something more sturdy if it doesn't come standard. For example, when I received my Meta Quest 2, it arrived with a flexible, elastic band, which slipped on my hair. I was able to upgrade to a heavier duty strap with an adjustable knob to keep the device secure on my head.

Adjusting the device properly on your face is also the key to seeing the content properly and clearly. Almost all devices have an option to adjust the lenses to match the user's IPD, or interpupillary distance. Typically, this process is done manually by sliding the lenses, or using a knob. However, newer devices, such as the PICO 4E, are introducing automatically adjusting lenses that require only a brief calibration once the headset is on. Making sure the device is sitting at the right place on your face is an important factor for clarity. Too low or too high, and you can't see through the lenses on a virtual reality device, or the content will be misaligned or out of view on an augmented reality wearable.

Comfort of not only the hardware but also the software contributes to the reduction or elimination of any sort of physical uneasiness.

Referred to as sim sickness, motion sickness, queasiness, etc., many individuals throughout the history of virtual reality have felt this type of discomfort. One of the biggest advancements in the elimination of motion sickness has been the introduction of positional tracking, which allows headsets to track six degrees of freedom (6DOF). These headsets not only track roll, pitch, and yaw, which can be demonstrated by how you move your neck when your body is stationary, but also movement sliding forward and back (surge), left and right (sway), and up and down (heave). With the addition of this tracking, all of the degrees that your body is processing and experiencing in the physical world can be mapped in the virtual world, creating a smoother and more comfortable experience.

In addition to the hardware advancements, comfort of the software also contributes to feeling good in the experience. Humans need a high enough frame rate (visual frames that are delivered from the device to your eyes, often measured per second) to feel like the experience is keeping up with what you would perceive comfortably in the physical world. The standard for most games is 72 frames per second (FPS) to feel comfortable; however, many devices support higher frame rates, and that is bound to continue to increase as technology advances. Other factors such as moving the visuals of the experience without physically moving the user or how the user moves throughout a world can contribute to comfort as well. We will review this in detail when we go over how to create an optimal user experience in the design and prototype chapters.

Appearance

We've all seen funny videos of people looking silly while using virtual reality in public. It's hard not to feel a little silly when you use it for the first time. But once you get into a headset, a lot of those feelings disappear. I've seen grown men who act tough and serious turn into a kid with no reservations after putting on a headset. I don't know if it's because you can't see the outside world anymore, or if it's just so engaging that you stop caring about what other people think. It is truly a magical experience. However, some of those viral videos could have been prevented if the equipment and space were set up properly and safely. There is no reason anyone should crash into their coffee table or punch a television off the wall.

Playspace Setup

It's important to set up your playspace, or the area in which you will utilize virtual reality with limited or no visibility to the physical world, based on the content you're using. Ideally, a 10-by-10 foot space is an excellent setup. Unfortunately, not everyone has that furniture-free open space in their conference room or living room. Most systems allow the user to draw their space while seeing through the headset with passthrough cameras. It's important to draw the most space you have available while being cautious of items around you such as furniture or decor. Based on what you draw, the system will show you a boundary grid in virtual reality when you get too close to that boundary to keep you safe. The device will also prompt you to set the location of the floor, which is a vital calibration step. If you don't accurately set your ground position, you may appear to be floating or be partially in the ground when you start using your experience. Typically, the device will be able to recognize the playspace setup you create between sessions so you will not need to perform these steps each time.

Advanced augmented reality headsets will map the space around you, so you do not need to manually set up a space. The beauty of augmented reality is that you will always see your world around you, so the playspace is wherever you are. While augmented reality isn't as immersive as virtual reality, it's a much easier and forgiving setup.

Hardware is always evolving, and with ever-changing options, the decision of which device to use can be daunting. Luckily, most experiences you produce will be compatible with future hardware releases with a minor amount of modification. Systems such as Open XR and Snapdragon Spaces are making it easier to build in cross compatibility across similar devices. Keep an eye on the horizon and anticipate change, but don't let that limit making a definitive choice today.

3

Finding a Use Case

A USE CASE is a documented solution that resolves a problem that an organization or audience may have. In our case, we want to define how extended reality can solve a problem, thus identifying the use case we will use throughout the production process. This use case will be the central driving force of design decisions throughout development and will provide a benchmark for measuring success upon deployment. Determining your use case is going to be a process in and of itself. Extended reality is excellent for entertainment, and it's also a powerful tool to leverage when solving problems. But what problem are we solving for? That is an integral question to ask before diving into use case definition. As we're looking for the problem to solve, think of what a good fit for extended reality might be. We already know that extended reality technology is great for simulating something grounded in real-world physics. We also understand that it has the potential to reach outside the realm of what's possible in the physical world. For training use cases, this could mean allowing an individual to practice their dangerous job in a safe space. It could also mean doing repetitions over and over again with unlimited digital materials, causing zero waste. No matter what the use case is, however, it's important to ensure it's applied to a problem worth solving—and furthermore, worth solving with augmented or virtual reality.

Problem Finding

A common process my team goes through is to write down problems, any problems, without thinking, "How will this fit into extended reality?" just yet. This can be done digitally or with sticky notes. When using sticky notes, write only one problem down per note. Sometimes it's good to do this in a group, and I always suggest this as a timed activity to put a little pressure on the situation. You can keep it more open-ended or give it a theme, which may be beneficial if you're trying to come up with problems that your company or department faces.

Some of the documented problems may not make sense or may feel silly. This is completely fine, if not encouraged, as it means that ideas were flowing without any worry of judgment. It often takes tens if not hundreds of bad ideas to come up with one good idea. The more you get on paper, the more opportunity you will have for a good idea to pop out. A stream of consciousness method can be effective as well, where you write down anything and everything that comes to mind, regardless of whether it fits the original prompt or not. Once all your ideas, concepts, problems, or thoughts have been written down, gather them together, and get ready to rank them. It's at this point you can share what you've written down if you're doing this in a group setting and combine duplicates if any.

Visualize in Extended Reality

After the problems are documented, it's important to identify which ones are possible to solve or elucidate with virtual or augmented reality. A component I like to start with is visuals, environments, and movement. Can you visualize your problem in 3D space? Does it enhance what you're trying to solve for? An example we can use for this exercise is leveraging virtual reality to learn how to use a table saw. I can visualize a 3D workshop setup with an interactive table saw. I don't necessarily have access to or own this equipment just yet, so having a virtual object to replicate this is helpful to me. Another element or question to consider is: What does this bring in a virtual reality experience that I can't do at home? For starters, I don't have the mess or material waste of practicing on a physical table saw in my house. My garage won't be covered in sawdust, and I don't have to buy practice wood either (or ruin my good wood that I intend to use for a home project). This experience would also allow me to practice something dangerous in a safe space to get comfortable with the process before performing the task in the physical world.

In an augmented reality alternative example, I am now ready to purchase a table saw, but I'm not sure which one will fit in my garage. Augmented reality would be a fantastic way to see what will fit in the space I have, and I can see the different options represented as 3D models in my space. I can flip through different machines, as well as visualize different colors and finishes to help me make my purchase. As an extension of the customer journey, I receive my table saw, but I don't understand how to operate it. Digital instructions and animations can also be overlaid onto my physical table saw to show me the order of operations and what each button does, helping me understand my new purchase. This digital content could also display safety and hazard content, helping me understand the importance of proper operation when using this product. Purchasing and use of a table saw would be a fantastic fit to leverage virtual or augmented reality, so the first step is complete: ensuring the problem can be solved with extended reality.

Audience Identification

After evaluating which problems can be solved with immersive technology, you'll likely be left with a much smaller handful of use case options. To continue narrowing them down, think about the audience you're trying to reach. In the table saw example, both the augmented and virtual reality solutions are more consumer-facing. I see these solutions aligned with a table saw manufacturer who wants to educate their consumers on how to use their products as well as ensure their shopping experience is successful. If this hypothetical table saw company wanted to train their own employees on how to ensure safety best practices while manufacturing their product, that would be a completely different audience. Instead of creating a consumer-focused experience, they should now focus on their internal employees and what will resonate with them. The goals of the project would change with this audience as well. Instead of selling more products and increasing consumer satisfaction, the new goals may be to create a safer work environment and set employees up for success.

We will typically document the intended audience by creating "user personas," which are fictional individuals that represent this audience. We assign them characteristics, demographic information, jobs, and even a profile photo so that we can visualize who we're creating for. Defining and understanding this audience prior to design and development will guide any questions you have throughout the process. Always think of your audience in any decision your production team makes.

Return on Investment

The final piece of determining the best use case is by calculating your return on investment (ROI). While in many cases ROI is going to be something that you can quantify, that may not always be true. For experiential or brand storytelling experiences, that value is harder to define. When producing learning content or sales experiences, it's much simpler. While you can estimate your ROI from the very beginning, it's important to note that this will not be proven until the project is deployed and you start to get data and feedback from the end users. There is still so much to be proven with the success of extended reality applications that it can be challenging to obtain buy-in with limited data. There are blessings and curses to entering an emerging technology field, and while it's exciting to be early to accomplish something new, it is even more important that we document its success (or challenges) throughout the process so that the industry can grow.

One point of measurement is cost savings on the reduction or elimination of travel. If the experience will allow people to perform training from their local office or home rather than traveling to a regional headquarters, document it. If it's a sales tool that can be deployed remotely or experienced by shipping out a headset to the prospect and it eliminates the need for a salesperson to travel, document that too. Travel can include flights, rental cars, hotels, meals, and a host of other expenses, which add up quickly. Scale those costs across an entire organization, and you're looking at substantial numbers that can be reallocated to the initial development of an extended reality solution and then saved in the future.

Another, more serious topic is safety. If there is a way to prepare employees and reduce the number of serious injuries and fatalities on a job site, that's a no-brainer. While the risk of human life is of the utmost importance, we're calculating ROI, which means we need hard numbers. As grim as it may be, how much money is spent on workers' compensation claims and lawsuits related to injuries or fatalities on the job? These payouts can range from tens of thousands of dollars into the millions depending on the severity of the case. Corporate insurance policies are also closely tied to safety records, so an improvement in safety can also affect those premium payments. If you're producing a training experience in an unsafe environment, the effect on safety is going to be the top measurable factor in calculating ROI.

Reduction in the use of physical materials has not only a cost effect but also an environmental effect. Depending on what is simulated with your extended reality solution, you are now using digital materials, reserving costly items that may have previously been misused or wasted. Even when using virtual reality as a design tool, we get unlimited paint or canvas, opening the possibilities of creativity. Going back to the table saw experience, I can practice with as many pieces of wood as I need until I get the hang of how to use the tool. I also don't have a floor full of sawdust and hours' worth of cleanup to do after I've had my fun. The same can be said for a manufacturing facility. Many times, the product that is being used for training purposes doesn't meet the quality standards required for sale and is therefore scrapped. Digital materials keep money in the bank and product out of landfills.

What if you could perform a task faster and better? While it's a phrase typically used at the gym, "sets and reps" applies here as well. My band director in middle school didn't say, "Practice makes perfect"; he proclaimed, "Practice makes permanent!" When you think about that, it's true. If you don't practice properly, all your effort will lead to permanent habits that may or may not be desirable. When you perform these sets and reps without guidance or oversight, you may be setting yourself up for failure. With a built-in guide or coach that is programmed to help you succeed and perform tasks correctly, sets and reps take on a much more substantial meaning. With this guidance, you could learn to perform a task more efficiently and with better quality, which could have tremendous financial impacts when scaled across an organization or to a large market. Measuring anticipated improvements to efficiency can ideally be added into the full calculation.

When using extended reality to visualize an environment that doesn't yet exist, design changes can be identified thoroughly before construction. Design reviews in 3D can be eye-opening, as you can see and identify issues that you may not notice on a 2D screen. If rework is a common budget item, consider how that could be reduced or eliminated if you are able to fully walk around and explore a space before it's built. There are countless additional opportunities to identify ROI that will be specific to your use cases. Think about the impact that your experience will have in all facets of deployment and what it could save you both in the short and long term.

Finalize Your Use Case

By following these three steps—identifying whether your problem can be solved with extended reality, defining your audience, and calculating your anticipated return on investment—one of your potential use cases should rise to the top. Once you've identified which is the best problem to solve, it's time to move onto defining the use case in more detail. There are many layers and components to creating an extended reality experience, so the more information that can go into this documentation, the better. We spoke earlier about audience being an important component to choosing hardware as well as calculating ROI. The audience is also an integral part of use case definition. The audience will define the direction, tone, and deployment of your use case. Is this something you're creating for an internal audience within a company? If so, leveraging proprietary processes and company information may be acceptable, if not encouraged. Is this a marketing or product assistance tool that will be shared with your entire consumer base? On that note, it's important to work with branding, marketing, and legal to ensure you're sharing the proper, approved content.

Before fully detailing your use case in a storyboard or user flow, it's important to speak to subject matter experts and do your research. More often than not, extended reality producers are not experts in the content they're producing, and on the other hand, subject matter experts are not familiar with the nuances and processes of producing extended reality content. It can be a beautiful partnership, especially when both sides of the coin want the project to succeed, and it all starts with a conversation. In the initial conversation with your subject matter expert, it's important to gain as much knowledge as possible about the actual process, environment, and experience you'll be re-creating. You want to become so knowledgeable about the content that you could leave the field of extended reality and move into their industry. It may seem as if I were joking, but I've had clients on multiple occasions in very specific trades say if my team members are ever looking for a weekend job, to head their way. Some important components to identify in these meetings include:

- What does the physical space around this process look like? How clean or dirty is it typically? Is there clutter in the space or is it organized?
- Is there specialty equipment that the user will perform their tasks on? Do you have diagrams or reference photos/videos that we can leverage?
- What do other people in this space wear? Is this apparel or wearable equipment required to complete the task at hand?
- What is the process that we want to replicate from start to finish? Are there any branching steps that can be taken? (i.e., are there opportunities for the user to make a choice or is it linear? If they can make choices, what are those outcomes?)
- What happens if the person completing the process makes a mistake? What are the consequences? Do you want to show these consequences in XR?

Once these questions are answered and well documented, the use case can be further defined in a storyboard. It's extremely important to get what's in your brain down on paper. Whatever you document will be leveraged as a guide for developers, artists, and finally, testers. This will be the start of your design documentation, which will expand throughout the project and serve as a single source of truth for the decisions made in the production process. Creating a simplified storyboard is a great place to start. Similar to a theater production or play, you'll want to outline what's happening in each scene in writing and when applicable, add a simple visual reference such as a sketch or picture. In a future step, we move on to creating what is called a visual storyboard, so this reference can be more rudimentary for now.

While you're outlining what the experience will be, take note of any interaction points throughout. It may be helpful to highlight those in a different color or bold the text. Each interaction is going to be implemented in the development process, and if this interaction doesn't exist yet in the deployment method you chose, someone is going to have to design how that interaction will occur. When speaking broadly about virtual reality, there are some common interactions that take place in almost every experience. Grab is likely the most common as it is the basis for almost any object someone interacts with in virtual reality. It's common in today's control mapping for users to use the "grip" button, or the button where the middle finger rests, to pick up and drop objects. However, this hasn't always been the case. When I started working in the industry, grip didn't exist on controllers, and we used the "trigger" button, or the button where the index finger rests. This was a tough transition to make for many, and some experiences still haven't made the shift, making it confusing for the user who is expecting to pick up items using grip.

Try to leverage as many of these common interactions without changing them so that it's easier for users to take their knowledge of how to use virtual reality across experiences. It's okay to introduce a new interaction if there is no standard or if there truly is a better way, but also remember to stick to basics when possible.

After the initial storyboard is written out, it's important to follow that with a user flow. The user flow will provide guidance on the outcomes of interactions or choices that may cause different paths for the user. Take the outline you've written in your storyboard, and put it into blocks. Every time there is an interaction or choice, those also become separate blocks. Use lines with arrows to show how those blocks are connected. Not everything needs to be linear either. There may be times when certain choices cause the user to revert back to a previous step or section in an experience. At the end of mapping this user flow out, it will likely look like a web, with many outcomes for all the interactions that you have thought of to this point. The user flow will also help you see where there may be gaps or discrepancies in your storyboard. Both of these documents are complementary to each other to ensure you have covered all your bases in the use case development process.

The final step in solidifying a use case is end user feedback. You've already been able to gather some great insights on the physical world process through discussions with subject matter experts, but going back to a small pool of your intended audience is a great step to take throughout all parts of the production process, even at the very beginning. You'll want to find people within this audience that have an open mind and the ability to see beyond what is created to date. Remember, you're only going to engage this audience with a concept, storyboard, and user flow. You must be the storyteller and champion of your idea but take feedback gracefully and understand that if your audience isn't on board with your concept, there may be a valid reason.

In these discussions, you'll want to introduce the concept, set the stage for the entire experience, and then give them an overview of the interactions and outcomes that will take place. The goal of this initial feedback session is to see whether your intended audience is receptive to the solution and sees an extended reality output as a good vehicle for the content. Questions you may want to ask in this session include:

- Does the overall premise of this experience make sense?
- Is there anything missing that you would expect?
- Does the flow of interactions and options align with how this would occur in the physical world?
- How could you see this use case expanding in the future?

The goal here is not to change your entire use case but to confirm that you're on the right track. This subset of your audience, if willing, would be great to continue speaking with throughout your process. As you have an initial prototype experience, alpha builds, beta releases, and finally the complete version, they will have seen it all and can serve as a sounding board in development. It's important, however, to express caution when engaging with an external audience who may not see the full picture or long-term goals of your project. Take feedback with a grain of salt and trust your own judgment as you're making decisions on what to act on and what to disregard.

Continually documenting and making revisions as you get these various streams of feedback is important. This needs to be legible and clear to serve as the link between a designer's mind and a developer's skills. With a solid foundation, including a well-defined use case, basic storyboard, and preliminary user flow, the true production process can begin. And while the design process may seem lengthy, it's critical to creating a good experience. The more detail and thought that goes into the design process, the less rework overall. You've laid the foundation to creating something great.

4

Reference

THE IMPORTANCE OF gathering great reference material cannot be understated if you're producing an experience that is grounded in the physical world. If the experience is more abstract, reference can still be important for inspiration; however, creativity can overtake the landscape freely. For learning and development or sales and marketing experiences, get as up close and personal as possible with the subject matter so that you can properly translate that into a digital representation. Ask for access—the worst someone can say is no.

There are several types of reference material that are necessary to capture. Some require a hands-on approach, while others can be provided by a third party. Regardless of who gathers the materials, there should be an organized method to store everything so that all members of the production team can have access. This can be done through a folder hierarchy in a shared file storage system such as SharePoint or through a collaborative wiki program such as Confluence. Designers, artists, developers, and testers may need to refer back to these materials throughout the process—both for initial production as well as quality assurance, so be as thorough as possible when collecting this information.

Most reference material we collect falls into four categories—visual, environmental, experiential, and auditory. Exploring these categories in the physical world, especially firsthand, will lead to a more authentic experience. There is some crossover between categories, and each category is not cut-and-dried for every use case. Expect to notice things outside the box as you're doing these intake sessions. Sometimes, the individuals who are providing reference are too close to the content and don't see all the details that you will see. An outside perspective is helpful and important for identifying all the small pieces that will come together to make a great experience. Cherish and take advantage of the freshness of your eyes because this will not last. The more familiar you get with the materials, the faster the blinders will go up and your eyes will start to glaze over at details that were once new.

Visual Reference

While some aspects of 3D content and extended reality can be more technically challenging, the visuals are almost always going to be the first element people notice. Visual reference not only includes the sights around a space but also the movement and flow of what's in that space. Photo and video reference are the most common when taking in the visuals of an environment or product. Another type of visual reference that is less commonly provided are 3D models. It's a luxury to receive these as part of the process and makes translating visuals into an extended reality medium much easier. Typically, these models are provided as engineering files, so they are much too heavy for one-to-one transfer. However, they do give us a digital starting point, which saves time. Unfortunately, these aren't always available, so we supplement with other visual media.

If you must rely on photos and videos, try to take them from all angles and in various lighting conditions. Notice where the lighting is coming from and capture the shadows on the objects. Photograph different components from far away for perspective and up close to get the nitty-gritty details. If you have access to a 360-degree camera, take advantage of it. While these images won't give you full detail, they will help your 3D artists know how everything in the space fits together. These 360-degree captures can also be helpful visuals for your detailed visual storyboard, which will come after all reference material has been collected.

After static spaces and objects are captured, turn your focus to movement. If you're in an outdoor space, how does the weather affect the world around you? Is there a light breeze that rustles the trees? Is there a cloud of dust when a vehicle drives by? If you're inside, focused on a piece of machinery, how does that move? What components are static and which ones should be animated in the experience? Take videos of everything, up close when possible and then bigger picture to capture how movement fits within a space. And if possible, bring an artist or animator into a reference capture session so that they may also see with their own eyes.

Environmental Reference

When I say "environmental," I don't mean the literal environment you see around you. I mean how the environment makes you feel and what little details add up to make it a full space, rather than objects cobbled together. What's the glue and substance behind the items you see? Take a moment to breathe in the space. Does it feel clean or dirty? Are there elements that you notice that would make the virtual environment more realistic if those pieces were brought into extended reality? While there are limited hardware options on the market for smelling in virtual reality, companies are working on it. Take that in too if you're able to do this reference intake session in person.

Aside from the physical attributes of a space, how does it make you feel emotionally? Is it scary, daunting, or intimidating? Is it friendly, jovial, or comforting? The elements you can create virtually can make an impact on this feeling when the audience is experiencing your content. They may be translated into extended reality in a different way, but the feelings that are evoked should persist in both the physical and digital representations of a space. Perhaps you feel happiness and comfort in a field of flowers in the physical world. Visual effects, including lightly floating particles and stylized art, can bring that feeling out, even if the experience isn't identical. Remember to note the abstract, not just what's in front of your eyes.

Experiential Reference

Interactivity is such an important element of virtual content, and inspiration can come from many sources. When replicating a process if the use case is training related, this reference may be the most pressing to capture properly as it needs an accurate representation when transferred to the digital experience. If the use case is more abstract, collecting reference material from multiple places and piecing it together can be more helpful. Regardless of what you're producing, getting experience with the product, service, practice, or subject matter is going to give you a lens of how your audience may experience something in virtual reality. Focusing on interactivity and what you experience in the physical world during these intake sessions is key.

When possible, ask if you can perform the process yourself in your reference environment. It's amazing the lengths people will go to when they share your vision of bringing content into an immersive platform. For example, if you are working with a table saw manufacturer, get clearance for a guided tour of the factory floor. Observe the process firsthand, and identify how people are moving and interacting with equipment when producing their product. Notice how tools and equipment are laid out in the space and if people must reach or take steps to switch between these items. A virtual reality experience may be confined to a smaller footprint, so knowing how much physical movement is required to perform a task is noteworthy as that information will go into spatial planning and design.

If you're producing a consumer-facing marketing product with augmented reality to help people with their purchasing process, go to a physical store where the product is sold. Observe consumers, and notice if they have any challenges making a decision to buy. If you hear, "I like the product but I don't know if it will fit in my home," then an interactive measurement and placement tool will be extremely helpful to them. If they don't know which color of an item to purchase, swapping out finishes in a virtual model is going to make an impact. All of these physical-world challenges will be converted into digital points of interaction, so understanding this feedback will help guide the design of the solution.

Auditory Reference

Last but certainly not least in my reference-gathering process is auditory reference, or the sounds, ambient noise, and voices within a space. Audio often gets treated as an afterthought in the production process, which is a shame. Extended reality is grounded in 3D space, which means we get to use something called spatial audio. This means that the audio elements can be placed with the 3D objects they represent. For example, if I have a bird in a tree within my experience, I can place the audio clip of the bird chirping in the tree, connected to the 3D object of the bird. When I am standing in my space, I hear the bird as if it were above me. If the bird—the 3D object—flies away, the audio clip moves with it, so it sounds like the bird is flying further away from me. There is an incredible feeling of presence when audio is done right in a spatial setting. This is why it's critical to pay attention to not only what sounds you're hearing in a reference space but also where they're coming from.

In general, capturing the ambient sound of a space is a great starting point. Record a clip while standing in the middle of a space for a solid few minutes. If it's captured with a good quality microphone, it might make it into the final implementation. From there, you can start to notice specific sounds coming from a particular machine, vehicle, animal, or object. Try to document these separately. They will likely be replaced with Foley—reproduced or licensed sound effects—but knowing what these elements should sound like will help with accuracy later.

When you're in a space with people, don't forget to pick up on their chatter. If the experience you're producing includes dialogue, listen for the tone of voice individuals are using when performing their tasks, giving instructions, or even just talking casually in that space. If it's a loud space, do they need to raise their voice to be heard? Little details and nuances of how they interact with each other will help with direction if you choose to use voice actors in your experience.

Virtual Viability

After gathering all these references, consider what you can actually bring into these digital spaces. Also consider what you want to bring in. Not everything is going to be a fit, and replicating entire worlds or product libraries can be daunting. I mentioned spatial planning earlier, which indicates how you want items and objects composed in your 3D space. Instead of delivering content on a piece of paper or a flat screen, you will be placing these objects with depth and dimension all around the user. Think of what you noticed in your intake sessions about the focal points and major interactions you want to include. Are these even possible in extended reality? Will adaptations need to occur to make this functional in a virtual world? This will be worked out fully in the design process, but it's good to think about when it's fresh in your mind.

Replicating the physical world can be challenging, especially with the processing limitations of today's technology. So, when you intake the reference material, make sure to take note of the most important elements. Think about what background elements can be combined or omitted. Of course, it's ideal to make the most realistic digital environments possible, but sometimes there are constraints, such as budget, timeline, or hardware, that won't allow you to include everything. Note these nice-to-have elements as "stretch goals." Prioritize the main features, and then go back and integrate stretch goals if you have time and bandwidth.

Next, think about what you observed in the physical world that couldn't be done. Is there a way to make it possible in the virtual representation of the space? If you can't expand the inner workings of a complex machine in front of you but you can do that in a 3D model in both virtual and augmented reality, would that be helpful? Are there elements of the space such as air flow or particles too small to see that can be visualized with animations or effects? Identify things you can't see or do as you gather reference, and make note if this would be an optimal fit for extended reality. These tiny or abstract elements can be visualized in extended reality in a way that can't be done in the physical world, helping your users understand concepts or content in a new way.

While most of your reference material should be gathered early in the design process, opportunities will arise, and the need for additional detail may come throughout production. And if you create an environment or design an interaction that doesn't feel right when compared to reference, don't be afraid to iterate and change course. Be communicative with your fully integrated project team as these challenges come up, and consider your audience throughout this process as well. What will resonate with them? Ensure the details you're focusing on have the biggest impact for the resource investment. Once reference material gathering is complete, it's time to move to full-on design. This is where you take all your research and use case development to date and apply it into full-fledged design documentation so that the artists and developers can bring it to life.

5

Design

ONCE YOU HAVE your pieces—use case, simplified storyboard, user flow, reference materials—you're ready to get into full-on design. At this point, I suggest you start developing a visual storyboard, which can include 3D elements as well as an extremely detailed written component. You can develop this yourself if you have the skills and resources or work with a specialized team of extended reality professionals to do so, which are outlined as resources in the production chapter. There aren't many design-specific tools out there for extended reality production, so using a combination of UI tools (such as Figma or Adobe XD) and 3D sketching or positioning tools (such as Open Brush and ShapesXR) will allow you to properly visualize the elements before putting them into production. Keeping all this information in an organized fashion and on a shared platform will ensure you and the rest of the production team are working off the same documentation. I cannot stress the importance of this enough. If these documents are floating around on people's desktops or decisions are made in the body of an email that only 50 percent of the team was copied on, it's going to cause problems. We use Confluence as a place to keep our documentation, but any collaborative wiki platform will serve this purpose.

Before adding any visuals, take your simplified storyboard and incorporate as much detail as you can from the reference material you gathered. Fill in all of the gaps of the outline. Describe the environment and all the required elements within it. If a character is wearing something special, what does it look like? Is there movement in the scene that's necessary to the story or content? What time of day is it? Is the user going to be outside or inside? What visual details are most important to focus on? As hardware limitations may require you to scrap visual elements, focus on what will make the biggest impact, and elaborate on that.

As you go through this level of detail, think back on the goals of the use case you have defined. Consider what a successful outcome looks like and keep that in mind as you're adding detail. What information did you learn in the reference process that can be incorporated into extended reality? Are there components that didn't exist in the physical world but are needed in the virtual experience? And alternatively, are there elements in the physical world that wouldn't be a fit in extended reality? If the goal was to expose people to learning content, ensure all the processes and educational material are incorporated in a way that aligns with that content. In that case, the material is not subjective, so careful attention to detail is crucial.

If you're creating an entertainment or gaming experience, creativity can flow throughout the design process. Leverage your imagination to fill in the details. One of my developers once told me to document anything I wanted—the sky was the limit. Once that was done, we worked together to pare down what was feasible technologically and created a final design from that working session. If you worry about what's possible during the creative process, it may limit the experience. The "what's possible" sessions are some of the most fun you'll have in the process, so make sure to take advantage of the creative freedom and truly define what you want if there were no limits. Leave it to the technical team to determine if something can or cannot be done, and then adjust the design documentation to fit within those limits while still maintaining your creative vision. More often than not, we find that there is a creative way to get most of the way to that vision, even if some concessions have to be made.

As you continue through this ideation and documentation process, remember to update the user flow as you go. This includes any interactions with reactions, as well as choice-based outcomes for any narrative elements. User flows can be created in any software that supports flow charts; however, one platform we like to use is called Miro. This tool provides flow chart templates and supports real-time collaboration features. Collaborative tools are always nice when working with a team, and it's nearly impossible to create and iterate on a paper-and-pen-based document (although whiteboarding and sketching out things physically is fine to start). Eventually, the user flow should balloon to something highly detailed that accounts for any outcome possible in the experience. This documentation will be integral for your developers as they set up interactions and flows within the development environment. The user flow will also be used during the testing and quality assurance process to ensure that the vision from the designer and production team has made its way all the way through to the deployed experience.

Continuing through the visual storyboarding process, you can sketch and draw things in 3D. Whether or not you're a 3D artist, sketching something spatially can be helpful. The easiest way to do so is to leverage a program such as Open Brush in virtual reality. It's a simple program to pick up, and I can attest to that as I'm not an artist! Thinking back to the first paragraph of Chapter 1, make sure you're drawing in true 3D, not just what you think 3D is. If you're incorporating a table, don't draw a flat version of the table to represent that object. Outstretch your arms, walk around as you sketch, and fill the space as it would be in the physical world. Planning spatially in virtual reality can work as a part of the augmented reality design process as well. In AR, you're still going to be placing digital objects in a 3D space, so the same practices can be leveraged here. Just focus on the object placement and goals of that, rather than the surrounding environment (which will be physical in the final implementation). Programs such as Open Brush often have an export feature, and while it may be rough, it could be highly beneficial for the art team to see how the world has been spatially planned to bring it to life.

Another spatial planning design tool, ShapesXR, allows you to create "slides" similar to a standard digital presentation. This feature can be highly effective when you're trying to tell a story or show a part of the user flow in a visual way. Planning the placement of things and how they progress throughout the experience is a great way to knock out the potentially subjective details prior to full-on production. It's going to be costly (in both time and other resources) if these details are implemented and there is a misunderstanding from the stakeholder or original visionary. Rework, especially in virtual reality, is difficult and frustrating. This is bound to happen at some point—subjective details and the review process are unavoidable—but if the first time a stakeholder is seeing the experience is at the very end, that can cause big problems. It's difficult to get a vision out of someone's head and into another medium. It's even harder when the people bringing the vision to life are not the visionaries themselves. Checking in and showing progress visually will lessen the headache of rework and lead to more positive collaboration throughout the process. That being said, not every detail needs to be run by every member of the team at all times. Sometimes it's best to get an interactive feature complete or finish a set of cohesive 3D models before asking for feedback. As hard as you try to document the design, there will be changes throughout as the experience evolves. Make sure when this happens, the design documentation is updated with the latest decisions. Design documentation should include (1) environmental descriptions and sketches, (2) specialty models and featured equipment, (3) characters and their features, (4) dialogue for characters or narrator, (5) animation for characters and objects, (6) user flow with all outcomes, (7) interactive elements and controls mapping, (8) sound effects and ambient audio, (9) developer directions for transitions and locomotion, (10) user interface, and (11) scoring matrix.

Environmental Descriptions and Sketches

Detail out the surrounding areas of interest in the experience. Environmental design will be most important for a virtual reality experience, as the technology is going to encompass the user's visual senses in full. Include location, time of day, time changes, and weather, if applicable. In addition to thinking of the physical elements of the environment, think also about how the environment should make the user feel—welcoming, serious, whimsical, scary, serene. Use those descriptive words to help you visualize the space and write out the smaller elements that bring those feelings out.

Sketch as much as you can in both 2D and 3D platforms. If you are an artist, work on several concepts for the style of the environment. Keep in mind the technological limitations of mobile virtual reality hardware, and produce your concepts with that in mind. Envision the environment with different lighting conditions. Think about how the user will move throughout the space, and ensure the flow of this space is conducive to the actions that will be performed.

Specialty Models and Featured Equipment

In addition to the general environment art assets, there are likely to be specialty models or featured equipment that will anchor the experience visually. Or, if you're producing an augmented reality experience, specialty models or overlays may be the only art you need to show, hence its importance. When you design this equipment, think about which components will need to be inter-active. Will parts of the model, often referred to as a mesh, need to function independently of others? If so, those models will need to be set up with something called sub-meshes, which allow sepa-ration of the components of a 3D asset. Document your expected movement of these specialty models as you work through design.

Do these specialty models need to be true to reality? If so, find out if there are other 3D models in an engineering format that already exist. Unless the model is already optimized for extended reality, there is going to be a conversion process to lower the polygon count (surfaces that make up a 3D model) and ensure materials are set up correctly. It's a tedious process, but there are some sys-tems that make that conversion simpler and more automated. As these assets are designed, produced, and optimized, ensure they match the style and theme of the rest of the environment so that they can be brought into a cohesive setting.

Characters and Their Features

Not all experiences will require characters, but when they do, document each character and their demographic and personality makeup individually. I find it helpful to give them all names, even if they aren't going to be named in the experience. It's a lot easier to think about them, their animations, and interactions if they have a name. Document the physical features, such as eth-nicity, skin tone, hair color, hairstyle, age, clothing, and any other element you want to specifically relay to the character artist.

Moving beyond looks, document the personality and underlying qualities of the character. If they are a background NPC (non-playable character), they may not need much of a personality, but if they are supposed to be emoting or showing a certain mood in the background, still write it down. If they are an NPC that the user will interact with, document their disposition, give them a role or purpose, and identify if there are any specific mannerisms they should possess. An instructor or guide might be designed with a more authoritative tone and a one-track personality. A character that the user interacts with as they move throughout their journey may show excitement, concern, frustration, or happiness. When interacting with a character in an immersive way, they seem more real, so take the time to bring them to life in your design.

Dialogue for Characters or Narrator

After your characters and their personalities have been designed, leverage that knowledge to add dialogue to progress the story or experience. This doesn't have to be tied to a visual character either. For example, a narrator can be represented as a disembodied voice, but it should still have its tone and purpose defined during the character design process. As you're defining the dialogue for your characters, write out directions when applicable. You may want some lines to come across as witty or charming. Alternatively, there are scenarios where the tone needs to be serious or even angry. This is especially important if you're using voice actors for these lines as you want to provide as much context as possible to help them in their recording sessions.

Match the dialogue you've written to your user flow to ensure you've accounted for every outcome of the scenarios that you've designed. In addition to character dialogue, also consider instructional text or additional detail that isn't necessarily spoken. For accessibility reasons, it's helpful to display all spoken text on panels within the experience. These panels can include other information, so use that to your advantage if there is a good reason.

Animation for Characters and Objects

Animation is a fantastic way to bring an experience to life. It also requires a great deal of forethought and setup to be executed properly. For characters, think of how you want them to move. Include that information in their individual character design. If a character is older, perhaps they move a little bit slower than the younger characters. If it fits the narrative, a character might be clumsy or have specific mannerisms that they express when they speak. There are free and paid libraries of standard character animations, such as Mixamo and Character Creator, with options such as walking, running, sitting, kicking. These serve as a wonderful starting point; however, it's likely that specialty animations will need to be produced. These can be animated by hand or produced with a motion capture system. Either way, reference to the design documentation will be crucial to bring these movements to life, so include detail and perhaps even a video of the action.

Objects, whether prepared to be overlaid onto the physical world or included as a virtual environment, may need to be animated as well. Animations are an excellent way to show the reaction of an object interaction. For example, if I push the right button, perhaps a conveyor belt turns on and starts moving product down the line. In addition to detailing the animation itself, also make sure to include the reason for the animation and if it's tied to an interaction. Other environmental movements may not be represented by animation but as particle effects. These effects are often used in digital content to represent motion such as water, weather, or fire.

User Flow with All Outcomes

The user flow keeps coming up, and that's for good reason. I consider it the most integral piece of your design documentation. That may sound like a bold statement, but without it, there is no structure to all the incredible design and detail that you have been documenting. It's the heart of the production process, and everything stems from it. There are commonalities across user flows, but different design platforms may use colors and shapes in a variety of ways. You can choose from some of the templates, or if creating from scratch, I suggest that you set a standard for yourself, create a key, and stick to that format (at least on a per project basis). For example, green may always mean a positive outcome and red signifies a negative one. Or a parallelogram always represents an interactive action, whereas a standard rectangle is just a simple step.

Using shapes and colors can help you see where there are gaps in the flow, as well as if a part of the diagram is too complex. Use arrows to show movement throughout the path, and remember that progression doesn't necessarily only go one direction. Incorrect actions could lead to the outcome of moving back a step or two (or even getting booted back to the start). After you think you're done, go through the flow in your mind, making choices and treating it like a "choose your own adventure" book. If something doesn't make sense, rework it or remove it. Get an outside perspective as well. Sometimes when we're too involved in something we miss out on obvious shortcomings. Again, every component of the experience will drive back to the user flow, so design it with intention.

Interactive Elements and Controls Mapping

As you've defined your objects and how they'll be animated, you will have also thought about which of these objects should be interactive with the user. In a virtual reality space, most (if not all) of the elements within reach of the user should be interactive. If they're not, the user may think the experience is broken or they could get frustrated. Interactivity can be as simple as the ability to pick an object up. If you can pick something up, you can place it, throw it, stack it, etc. Not everything has to be fully functional, but basic interactivity leads to a higher degree of immersion. For more complex objects, the ability to open and close hatches, push buttons, pull levers, etc., could be necessary.

Make sure to document not only how the element is expected to be interacted with but also what controls you want to be used to initiate this interaction. Most controllers use "grab," or the button associated with the user's middle finger, to pick up and drop items. This has become standard and feels natural; however, if it's not a fit for the experience, change it. Hand tracking is also becoming more prevalent, so gesture control associated with these interactive objects could be a fit, depending on the deployment method. There are even headsets on the horizon that won't include controllers as a part of their standard. Ensure that the controls mapping that you document aligns with the hardware that you will be using for production.

Sound Effects and Ambient Audio

The audio aspect of any digital experience has the power to influence the user in such an immersive way that we need to plan for it from the very beginning. What's even more exciting is that any 3D deployment, both virtual and augmented reality, can leverage spatial audio to allow sound effects or dialogue to come from the origin of the sound. Go back to the audio you documented in your reference-gathering stage and remember what was associated with each object. All of these will need to be incorporated into the project as a separate audio file, which means they will need to be sourced independently. If they are associated with a specific object, make sure to document that so that whoever is doing the technical integration can map it spatially.

In addition to individual sound effects, ambient audio plays a large role in creating an immersive space. You can license or create new ambient audio, or you can leverage the recordings you took in the reference period. If you're designing a manufacturing facility, why not leverage some of the humming and white noise you captured on the floor. If you're designing a quiet space, it's still good to include some sort of ambient sound (even if faint). Silence is loud and it's noticeable, even distracting, if you don't include any ambient audio.

Developer Directions for Transitions and Locomotion

When developer directions are written out, the documentation starts to look like a film or theater production. This includes detail regarding any fades, jumps, or scene transitions. Remember, physically moving the user can be jarring and uncomfortable (or even nauseating if done incorrectly), so take great care and consideration when planning these transitions. In a narrative experience, fades or cuts may be necessary. By using a fade, this quick transition method mentally prepares the user and lets him or her know that something is happening, so they are not just "blipped" into the next scene or location.

Planning your locomotion method is great to document as well. Choosing to support teleportation, smooth locomotion, or even a more creative method of user movement should be designed at the start. Think about accessibility during this time, as well as the amount of space you expect your user to have. If most of your users will only have a six-by-six-foot space, make sure you design ways for them to get to things that are out of that boundary.

Define if you're going to support a seated and standing experience, or if you will only design for one. If you will allow the user to move freely throughout the space, include it in the developer directions. Alternatively, you can restrict teleportation to what we call "hotspots," which are commonly placed at points of interest and areas with a lot of interactivity. If you're leveraging hotspots, document where you want these to be on a map of the environment for developer implementation.

User Interface

Your user interface (UI) design will be crucial for augmented reality as most of these experiences are driven by menus and tools on a 2D screen. As spatial computing wearables become more common in the augmented reality space, there may be more crossover with how UI is incorporated in virtual reality. For augmented reality, you can get inspiration from how you interact with menus on your phone or tablet. There is a need to find the object or overlay, place it, and interact with it. Think of iconography and what tools you'll need to activate throughout the experience, and design what you'd like those to look like. Also consider the amount of clicks or actions it takes to get to the root of the interactive experience. Ensure it's clear how to move from one feature to the next.

User interface design for virtual reality still has commonalities with 2D menus, but there are opportunities for more creative solutions in a 3D space. The biggest UI elements in a virtual reality experience are going to be the start/end/settings menus and any panels with text pop-ups throughout the experience. Ideally, the experience itself will fully leverage interactivity and spatial immersion that makes virtual reality what it is. Try to minimally incorporate 2D UI elements so as not to break that immersion.

Scoring Matrix

Most games or training experiences will require a scoring matrix to determine whether an outcome or action is successful or not. This is something that can be driven by the user flow but will need a bit of additional documentation to implement. Take all the outcomes from your user flow and put them in a chart. Then add a column to either associate a pass/fail outcome or a numerical score associated with that line item. You may not want to have every action affect the score, so only include what's needed.

If there is content associated with the success or failure of any of the scoring actions, document that as well. This text can be displayed at the end of the experience on a scoring panel or can be represented by a voice-over as well either at the end or immediately after that action has taken place. Scoring matrices can become overwhelming and complex, so use tools such as Excel to your advantage, and set up filters to help you sort through the outcomes and formulas to calculate your scoring.

At this point, you should have quite impressive and effective design documentation. The amount of content can be overwhelming, and depending on the complexity of your content, the next steps could lead into a phased approach or a full-on development effort. This will all depend on the timeline and budget you must work within. Regardless of which path you choose, designing the big picture in detail and then going back to define the road map and production schedule will allow the long-term vision to carry through.

6

User Experience

PRESENCE IS DIFFICULT to describe as it's completely subjective, but it's imperative to any extended reality experience. The meaning of presence, as defined earlier, is the feeling that you get when you're in an immersive experience. You lose track of reality, and digital and physical worlds collide, which leads you to believe in what you're seeing. While this term is more prevalent in a virtual reality environment where your visual senses are completely occluded from the physical word, I believe that it can also be applied to an augmented reality experience. If the content you're visualizing overlaid onto the physical world is seamless in its spatial tracking and interactivity, it can feel like a piece of the world. That's what we're looking for—the blend of the digital and physical and the feeling of belief.

If you're ever stumped as to why something in your experience doesn't feel right, think back to the idea of presence, and identify if it exists within your content. If the answer is no, ponder what is missing. Is the audio balanced? Are there digital distractions that make you feel like it's not real? If an object is glitching out or the experience is too quiet, those problems can distract the user, and they won't fully believe they are a part of your world.

Immersion refers to the objective elements that make up the experience, whether it's the hardware, software, or interactive components. An example of an immersive element is the implementation of an interactive object, such as throwing a ball. If the picking up and throwing mechanics of this ball-throwing action feel correct, the interactivity is immersive. If there is a delay, glitch, or the interactivity is unexpectedly difficult, this can break immersion for the user, which tarnishes the experience.

Presence and immersion are often used interchangeably, although they represent different interpretations of an experience—the subjective and the objective. When thinking of the user experience and how to craft it, keep these elements of presence and immersion at the forefront. Contributing factors of presence and immersion include (1) visuals, (2) audio, (3) haptics, (4) physical comfort, (5) mental comfort, (6) social comfort, (7) controls tutorial, (8) teaching interactivity, and (9) guardrails.

Visuals

Make sure that whatever you incorporate in a visual sense is believable. That does not mean it has to be rooted in the reality we know and observe in daily life or even of this world. However, the details of the visuals must come together in a way that is cohesive and structured. Even an abstract environment can draw out a sense of believability. But if you incorporate random elements with no common theme or framework, the user will be confused, which pulls their attention to that confusion and removes their mental focus from the experience.

In addition to the objects themselves, the stylization of the visuals needs to be consistent. Determining an art style early on and ensuring all the visual elements align with this style will not only help guide the production process but also add to the believability of the digital world. Choosing a style that fits within the goals of the use case leads to a stronger user experience. For a training application, visuals that are true to the physical world representation are integral to an effective learning experience. For an experiential marketing application, a whimsical or futuristic art style with creative flare will lead to the eye-catching goals of the use case.

Audio

We've already touched on the importance of audio in design, but the proper balance and implementation of audio can have a make-or-break impact when it comes to presence and immersion. The biggest impact audio can have on presence is when it doesn't exist at all. Think about the world as we know it and how much our sense of hearing has an impact on daily happenings. Even as I'm writing this I can hear the light hum of my air conditioning, my dogs eating their breakfast, the swishing of my dishwasher running, and the clicking of my keyboard. There are probably dozens of other sounds that don't come to the front of my mind that exist around me and make up my feeling of presence in the physical world. Trying to think of every little detail can be overwhelming, and sometimes unrealistic, but incorporating background noise and sound effects for key items at the very least has a major effect.

Audio quality also has a tremendous significance when producing an immersive experience. This doesn't necessarily mean sourcing audio with the highest quality, although you should aim for the best technical quality you can achieve. Having consistency in the quality of your sound files and voice-over clips is most important. In many of our projects, we source voice-overs from various actors who do their recordings in their own studios on their own equipment. Typically, any professional voice actor will have high-quality equipment, but it can vary. If you insert conversational voice-over clips between characters and the audio quality doesn't match, it will sound unnatural and pull the user out of their feeling of presence. It won't be believable. The same goes for sound effects. Ensure that they sound good together and that the quality matches.

Haptics

Haptic feedback is directly correlated to the sense of touch and how it's incorporated into a digital experience. Something that you would be familiar with in a day-to-day implementation are the vibrations of your phone while playing a game (or even receiving a call or notification). When we talk about haptic feedback, or haptics, in relation to extended reality, there is an ever-expanding suite of options available. In the simplest sense, the controllers have the ability to vibrate and pulse, which when programmed correctly can make the user feel like they're picking up an object or moving their hand through a series of objects. This is technically implemented with colliders, which are a way that the program tracks when objects intersect with each other. This small amount of feedback gives the user a more tactile way to know they've performed an action, amplifying immersion.

In addition to controllers, haptic devices specifically designed for extended reality are constantly in production. One type of haptic device on the market is virtual reality gloves, some of which restrict movement of the fingers to make you feel like you're holding onto an object. Others leverage air pressure in tiny nodes to add pressure to different points of the hand, granting the user a similar feeling of precise touch feedback. Vibrating vests or even full-body suits are in development, which make the user feel like they have been hit or are intersecting with objects outside of just their hands. Even omnidirectional treadmills, which allow the user to walk or run infinitely in any direction, have seen traction. Location-based entertainment deployments have had the greatest success in using additional haptic devices because the experience is set up in a controlled and moderated area. Having a full haptic setup in someone's home is still a way off, but depending on the use case, leveraging multiple haptic devices can lead to a heightened sense of presence and an exceptional level of immersion.

A positive user experience relies on comfort significantly. Emerging technology is scary to people who haven't used it before, and with fear of the unknown comes doubt. To reduce that doubt, it's imperative that the user feel comfortable physically, mentally, and socially. Educating your audience can be a great way to ease some of these unknowns and ensure they embark on a comfortable and effective experience. As you consider how the experience will be deployed, work these three comfort factors into your communications strategy.

Physical Comfort

The biggest detriment to virtual reality technology adoption has been its historical tendency to make people motion sick (note: "historical" meaning it's mostly in the past). This issue has improved significantly over recent years, and most individuals don't experience it at all. There were several contributing factors to this discomfort that have been eliminated or reduced as technology has advanced, the largest being the introduction of six degrees of freedom, or 6DOF, tracking. The hardware's ability to track micromovements of not only our heads but also our position as we walk around and reach high or squat low makes our physical movements perfectly match the visuals we see. This connection between vision and physical movement means we don't feel sick! The other significant contributing factor to motion sickness is low frame rate, which is typically indicative of an unoptimized environment. As mentioned in the design chapter, it's imperative that the experience is optimized for the hardware on which it is going to be deployed.

Hardware comfort is another contributor to physical comfort. Ensuring that a head-mounted display or wearable is fitted properly will reduce the potential of headaches. In addition to the head strap, eye positioning also affects the user experience and leads to eye fatigue if not set up correctly. Most devices have adjustable lenses, so ensure users are fitting the lenses to the distance between their pupils, referred to as interpupillary distance (IPD), before starting an experience.

Mental Comfort

This fear of the unknown that we as humans feel has a big impact on mental comfort. We want users to feel secure in their surroundings and understand how to use the application so they can go into it confidently and have a positive experience. A great tutorial at the start of any extended reality product can lead to confidence, eliminate frustrations, and ensure that the user is focused on the impact of the content and not its shortcomings.

When virtual reality was becoming popular, many of my peers thought it was hilarious to scare their friends in their first experience of the technology. Whether it was a rollercoaster or a haunted house, this was the type of experience people wanted to watch their friends muddle through. I suppose this was to see the reactions on their faces, which could be entertaining to an extent, but this ended up turning many people off to the technology in their first impression because it deliberately induced discomfort. If the intention of the experience is to evoke intense emotions from the user, it's fine to incorporate uncomfortable elements such as heights and suspense, but these elements should be reserved for appropriate use cases. You do not want to unintentionally turn off a user because they are startled by or uncomfortable with the content. When introducing someone to extended reality for the first time, choose content that shows the incredible immersive power of the technology, not just a jump scare to get a short-lived laugh.

Social Comfort

An unfortunate fact of the online world is that there is the potential for harassment in any public (or even private) space. When using social media on your phone, harassment is commonly linked to harsh comments on a post. Moving one step further into computer or console gaming, hurtful conversation over audio can feel like an even more personal attack. But adding the layer of immersive technology and the presence of avatars in a shared 3D space makes harassment feel physical. It's imperative that for any multiplayer experience, social comfort is taken into great consideration.

Common moderation tools include the ability to mute and block other players, which for any public deployment should be implemented at the bare minimum. The concept of the virtual personal space bubble is becoming more and more popular in both public social applications and private multiplayer experiences. It feels strange if someone gets too close to you in the physical world as well as a digital one, so turning off the 3D model of the avatar when it gets too close is a common way to alleviate that discomfort. A social experience in extended reality can be incredibly powerful and connects us with people from anywhere in the world. Ensuring the user has the appropriate tools to control their social space will only increase their comfort level and keep them coming back for more.

Comfort leads to adoption. We can only get so much traction out of the "wow factor" of extended reality technology. There needs to be substance and ease of use behind the content to drive adoption and draw our audience into what we are producing. Tutorials can make or break an experience, especially when there are uncommon interactions or the user is new to the technology. Make sure to incorporate educational and tutorial content into your design to ensure a frustration-free user experience.

Controls Tutorial

A controls tutorial will teach your users about what buttons or triggers serve which purpose within your application. This is typically one of the first things that the user will see when starting an experience, and since you won't necessarily know their knowledge level with immersive technology, you must even account for a menu tutorial. Thanks to computers and screens, point-and-click is a common way to initiate any digital experience, but that action may be associated with a button or trigger on a controller, so don't leave it out.

Tutorials should be designed to wait and confirm that the user fully understands what you're trying to teach them. In the case that you're teaching them how to pick up an object, do not progress to the next step until they have successfully picked up that object. To enhance the tutorial, you can add failure prompts so that if they do perform the task incorrectly, the system explains why and then prompts them to try again with the correct instruction repeated. Try to include the core functionality as a part of the initial tutorial. If you don't want to spoil certain features or functionality, you can also insert mini tutorials throughout the experience to keep things clear but maintain the surprise. Tutorials are typically programmed as a requirement for the first-time user experience; however, there should always be a way to access the tutorial upon future launches. Inserting a "skip" option at the beginning of the tutorial can also be convenient for individuals who are familiar with extended reality.

Teaching Interactivity

When teaching a new user how to interact with certain objects, visually showing the steps may help with clarification. In addition to having a narrator or NPC explain what to do, show models of the interactable objects with highlighted components to get the point across. One interactivity type that seems to challenge newcomers is as simple as pushing a button. Pushing a button might not seem complicated, and when I ask you to do so in the physical world, you likely outstretch your index finger and push the button. In virtual reality, I've asked people to do that same action, and they outstretch their index finger and stop. They don't propel their finger forward to intersect with the button and then seem confused as to why the button wasn't pushed. As opposed to 2D digital media, extended reality leverages 3D space, and that final step of reaching out and actually pushing the button completes the interaction.

Sometimes this can feel as if we're overexplaining how to do some of the simplest of tasks, but sometimes the unfamiliarity warrants that over-explanation. Another 3D interaction that has caused challenges is reaching for something behind the back. Using a physical representation of a backpack for inventory is starting to be a common interaction. Once people get it, they understand it, but it takes a little extra explanation to indicate that the user must physically reach over their shoulder or behind their back, leverage the grip button, and pull the object in front of them to see it. We've used the same interaction to pull a safety lanyard out from the back of a harness. Actions such as these will require specific interactivity instruction to ensure the user can clearly perform the task. If the interactive element is more complex than "grab," add a custom section to cover it in the tutorial design.

Guardrails

When someone is unsure of what to do next, which is bound to happen especially with a larger audience, ensure you have incorporated what I call "guardrails." The simplest guardrail you can bake into the design is triggering repeated tutorial text or voice-over associated with a period of inactivity. This period of inactivity isn't necessarily consistent across use cases, so you can gauge this based on how it feels in your experience. If a user is supposed to be performing some sort of interaction and hasn't done anything, they're likely thinking about it or stuck. After 5–10 seconds of thinking, they may not even remember the prompt to begin with. It's after this period of thinking time that we like to trigger the initial prompt again. This should happen in your starting tutorial but can also be implemented throughout the experience in applicable places.

Another guardrail is directly affiliated with complex or uncommon interactions. Some experiences may be 30 minutes, 2 hours, 40 hours long. The user won't always remember the specific elements of the tutorial, especially if they don't perform these specialized interactive actions from the start. Inserting additional clarifying information at times like these can provide hints to the user so that they don't experience a point of frustration. These hints can be naturally incorporated into the experience through character guidance or can be more obvious with signs or visual indicators reminding the user how to interact.

When you're testing the experience, make sure to note where the potential points of frustration for new users lie. Then find ways to update the tutorials or add guardrails for a smooth experience. A user is not always going to have another person on "the outside" to guide them through, so incorporate as much digital guidance as you can.

Reducing frustration is important, but once all the interactions are clear and the user understands the goals of what they're trying to achieve, there are other questions to ask regarding user experience. Is it interesting? Is it enjoyable? Is it effective? Is it fun? Not all content is designed to be fun, but the root of these questions lies in confirming that this is an experience in which people will find value. If you produce the most technically advanced interactions or the highest fidelity artwork but it isn't enjoyable or effective to the user, you have failed. The answers to these questions are subjective, and they are determined by the user.

We'll go into detail about testing in Chapter 10, but it's important to touch on the impact of testing and iteration based on user experience feedback. You don't want to take every little piece of feedback from your users as fact. If it's a bug or issue with the functionality, that is worth consideration every time. However, the more subjective feedback, while important to collect and examine, can be taken with a grain of salt. Use this feedback to your advantage as you're likely much too close to the product design to see it clearly anymore. What you thought was a fantastic idea may not have the outcome you expected in implementation. Leverage this fresh perspective to shape out an impeccable user experience as you continue production.

Some of the suggestions or features that come out of this feedback may be unrealistic to implement based on time or budgetary constraints. Use your own judgment and filter system to determine which user feedback you act on and create priorities. This will come into play as you build out your production plan, work through scheduling, and identify your deployment strategy. Moving from design to production is an exciting and exhilarating transition as you get to see the fruits of your labor start to come to life.

7

Production

THE ENTIRETY OF this book covers the extended reality production process—design, art, development, testing, deployment. However, there are some foundational building blocks that fall into a producer or project manager's responsibilities. Like the user flow, these building blocks have the potential to shift and expand throughout production, but be cautious as that occurs. Stay close to the original plan unless there is a significant reason to deviate. When drafting your original plan or detailing the scope of the work you're producing, keep your limitations in mind. Whether they are budgetary constraints, deadlines, or content requirements, these limitations will help you in the end and ensure your scope doesn't balloon.

The project scope should be simply defined before the start of design documentation. This will typically occur in your use case development phase. Whether you're producing an extended reality application for a customer or for your own internal organization (or for independent deployment), you will need enough information to put together a schedule and estimate resources.

Content

You won't be able to write out the project scope without knowing what content you want to include. For an augmented reality experience, what models or filters are you going to include? Identify how many assets you anticipate producing and at what fidelity or complexity these assets are. Understand that if you're working from someone else's 3D files, there will likely be an optimization process to get these files ready for an extended reality deployment. Document whether these models include moving parts or animated components. Each of these sections of content development will require resource estimation, so try to be thorough when thinking through quantity.

In addition to art content, there will also be either narrative or subject matter–driven content. This information will help you document scope as it pertains to development. Is the experience going to be broken out into sections or modules? Is this going to be deployed as one linear path or is there a branching choice-based narrative? These are questions that should also be asked as you're building out your user flow, so use that piece of design documentation to your advantage when documenting the scope and goals of the content. Focus on the expected length of the experience, and account for that. If you're producing this with a client or review board, record how many rounds of revisions you will include as a part of the production process. The goal here is to account for quantitative content parameters in addition to the subject matter that you will cover.

Integrations

In addition to the content itself, understanding what systems you'll need to integrate with can have an impact on the scope, timeline, and budget. Most virtual reality experiences leverage a common framework called OpenXR. Augmented reality experiences may stand-alone in their own application, or content may be integrated in a platform such as 8th Wall, or SparkAR. There is an abundance of code plug-ins that offer pre-developed functionality, which a developer can incorporate into a development engine, such as Unity or Unreal. These take integration to fully function as well. Write down any third-party solution that you believe your team will need to leverage during development. And if the exact solution is unknown, still documenting that one is needed.

Analytics are another common component of an extended reality experience that require integration with an outside system. Whether the analytic data is being tracked on your own server or a third-party solution, linking that information from your program to that platform is an integration task. Some companies may want the data to integrate directly into their learning management system (LMS). While there are learning data standards such as xAPI and SCORM, these still require setup to get the systems "talking" to each other. Any integration, including analytics, should be documented as part of the project scope.

Deployment Method

Determining your deployment method at the start will reduce headaches when the product is closer to its release date. The first step is answering whether your content and objectives are better for augmented reality or virtual reality. Hopefully, you determined this during your use case development phase. If not, remember to identify whether it's content that should overlay onto the physical world (AR) or content that should fully surround and immerse the user (VR). Once that is determined, identify the specific hardware you will deploy on. This can depend based on your audience goals, but at the very least decide whether you'll release on mobile handheld devices, wearables, stand-alone mobile headsets, or computer-based systems.

How will your content get on the device you've chosen? You can release something publicly for a wide audience leveraging a consumer-facing store. Another option for a private but managed release is to leverage a mobile device management system, such as ArborXR. And you can always sideload or directly download content on devices one at a time, although that is not recommended as a long-term solution. There may be costs or technical requirements associated with your distribution method (especially if it's public), so include these details as part of your scope documentation.

Once your full scope of work is defined, it's time to start scheduling. This may not always be the case, but extended reality is still referred to as the Wild West, meaning there are innumerable elements yet to be standardized. That presents tremendous opportunities for people interested in exploring and producing in the industry but that also indicates that there are a significant number of unknowns. We can look at a simple website or a digital advertisement and say, due to historical data, that it should always take X number of weeks or months to produce. The same cannot be said for most extended reality work (especially custom development), as new, undefined features are bound to pop up throughout the development cycle.

Hard Deadlines

In my world, which heavily focuses on enterprise and business customers, there are often hard deadlines. And I mean hard. We once produced an experience for the Super Bowl, and that event wasn't going to be delayed for us—even though we were implementing several experimental features for the first time. Knowing what you're up against will help you retain focus and structure as you're developing your schedule. It will also help you prioritize. Most deadlines will not be quite as intimidating as "The Big Game," but if you're producing something for a training schedule or a specific event, start with that date from the beginning and develop your schedule backward.

Anything with an experimental aspect to it has the potential of unexpected challenges or delays, so think about your absolutes from the start. What are the elements that are absolutely required no matter what? Focus on these priorities in your schedule. Ensure you knock these elements out first before going back and filling in the details. If we identify an exciting and immersive feature that isn't necessarily in scope, we tag that as a stretch goal so that the team has something to look forward to if they get the priorities and required scope delivered early. Build the framework of your schedule around the requirements and priorities, working backward from those hard deadlines. Bake in double the amount of testing time you think you'll need, especially when working with a large organization. The review process is time-consuming, and while many of the deliverables are objective, there is also that subjective element of presence that will need to be reviewed.

Resources

The resources you have access to significantly affect production speed. Consider your resources—will you have your own team creating this? Are you leveraging a combination of contractors and members of your own team? Are you going to outsource this completely to a third party? Are you a solo developer making your own project? Identify the skills you already have, and document what components you need. Keep ramp-up time in mind if you are working with people outside of your existing resources, as project onboarding takes time.

When thinking about the type of resources leveraged in extended reality production, I typically include:

- **Production manager:** This individual manages the schedule and expectations of the content. They help with task management and communicate between the stakeholders and the production team.

- **Designer:** The designer gathers reference materials and produces all design documentation. They also ensure throughout the project that the original vision is carried out in development.
- **3D artist:** Artists will produce and/or optimize 3D models, as well as create environments, animations, and effects. They will also define the art style and ensure its consistency.
- **Developer:** Developers are responsible for the technical implementation of all content assets. They will implement interactivity and content as indicated in the design documentation and manage builds for testing and deployment.
- **Quality assurance/testing:** Testing personnel will review each build for accuracy compared to the design documentation. They will identify bugs and issues as they perform testing and confirm those issues have been resolved before release.

Determine how many of each of these resources you will need for your project. If you don't have background knowledge of their area of expertise, work directly with them to estimate the time, and plan it into a schedule. There are also certain components that rely on others to start work. For example, developers can't prototype the experience without design documentation or user flow. Consider how these different resources will stack within the schedule and what parts are reliant on others.

Experimental Features

Some experimental features are planned, and others are inspired by the production process as the user experience shapes out. When working in an emerging technology field, there is abundant opportunity to create something new. Bake in time for these features into your schedule, and know that design and implementation may take two to three times the amount of time you may expect. One of my augmented reality projects included a new tool that allowed the user to map out their space and then see when the product they were visualizing intersected with the mapped space. The content featured large construction equipment, so the intent was to leverage this tool to see whether the object would fit in the space or reach the work area. The implementation seemed simple enough on paper, but as we worked through the user experience and how individuals would access this tool, it became more complex, resulting in over twice the original planned development time. Don't underestimate experimental features, even if they seem simple on the surface.

Some experimental features will not make the final cut, and that's okay. Ensure your team doesn't get distracted from the overarching goals of the experience by new ideas that come up throughout the development process. If a feature is not part of the original scope, or it's not required, consider labeling it as a stretch goal, or a goal that you will dedicate resources to if everything else is complete. And whether the experimental feature was part of the required scope or not, ensure that it fits within the greater vision. Don't be afraid to scrap something if it's not working for the betterment of the user experience. People tend to be victims of the sunk cost fallacy, where they will not drop an idea because of the time and effort that was invested into it in the first place. While it's great to attack experimental features with optimism, it's also okay to change course if something isn't working out.

As you work through your time estimation, you can leverage task management tools to create high-level schedules and then incorporate details as you have more information. Be cautious of how much time you spend on this before a project being green-lit. Building a full project schedule with details can be extremely time-consuming, so focus on the high-level buckets of tasks and get a general schedule outlined first.

Scheduling Tools

Teams work in different ways, from agile to scrum to waterfall to kanban. There is no right answer in production, only what works right for you. I won't go too much into the details of scheduling tools as they are nuanced and each one likely has its own book. My team uses Jira, but any task-tracking tool should work if configured for your needs. Determine how long each of your development cycles (often referred to as sprints) will be. Some fast-paced projects move through cycles every two weeks. When we have a smaller or slower project, we'll typically align each bucket of tasks within a calendar month. Build out these cycles starting from your hard deadline so that you can ensure scope fits within these confines of time.

Once you have your cycles defined, start incorporating high-level tasks within each of them. Remember that many tasks are reliant on other tasks to be completed first, so as you're scheduling, don't plan those to run concurrently. Many tools will have the option to link tasks, and even mark them as "blocked by" another task so that the whole team understands where these dependencies lie. Build as much of your scope into these high-level tasks so your team can determine what is realistic. If it seems as if there is too much to accomplish within the schedule, take some of your "nice to have" features and identify them as stretch goals. Clarifying scheduling expectations from the start and verifying deadlines up front will reduce frustration later on in the development process.

Prioritization

Prioritization is incredibly important, whether you're working as an individual or with a team. As you think through priorities (especially when you have too many features for your schedule), document which features are going to offer the biggest bang for your buck or make the biggest impact to the overall experience. And don't only think about prioritization regarding individual tasks, be self-aware with how you discuss features in creative sessions. Several years ago, we started developing a virtual reality gardening game. My team spent multiple meetings over the course of days determining how a shovel was going to work in the first month of the project. It would have been much faster to make a decision, test it out, see how it felt, and then finalize the implementation. Instead, we hypothesized, wasted time, and ultimately had to make changes down the line anyway. If we had properly prioritized, we would have realized that setting up the game's base systems was more integral at the time than finalizing the shovel interaction, especially so early on in development.

Moving beyond your initial production schedule, reprioritization may come into play. When you're closer to deployment, do a recalibration with your priorities, and change them if necessary. It's at this point you'll also start to get an influx of bug reporting from your testing team. Those issues will need to be prioritized as well as incorporated into your later development cycles (which should have time reserved for these impending bug fixes). To help prioritize these unexpected development tasks, note which bugs are most noticeable. Yes, we want to make the best product we can, but priorities will help guide the team into making tough decisions, especially when you're down to the wire.

The production role focuses significantly on organization and planning, but there is also the task of communicating between all parties involved. When there is subjectivity, that introduces opinion, and with opinion there can be conflict. Promoting a healthy team dynamic and a space for collaborative thought is a soft skill that is often underestimated. I feel that any good production manager will have the skills to serve as the glue that holds the team together and sometimes as a mediator. Mediation, respect, and a healthy use of the word "no," are elements that I've incorporated into my team dynamic, and hopefully you can do the same.

Mediation

When the machine is running smoothly, the work is productive. Communication is the key to keeping an extended reality production operating efficiently and collaboratively. The dynamic between the design, art, and development teams can clash, as does any role with creative input. This doesn't mean that the team isn't functioning properly; it means there is healthy discourse. But when it gets more heated, it may be time to mediate.

Whenever this happens, in my experience, it's almost always been a communication issue. Developers or artists have misinterpreted the design documentation. Developers haven't received files in a preferred format from artists. Artists feel their work isn't appreciated. Bringing all feuding parties together and talking about it can diffuse the situation and provide clarity on expectations within production. If you find yourself as the mediator, try to facilitate the conversation. Don't let one person steamroll the others. Speak up and call on someone who is being exceptionally quiet. Make sure all voices are heard, and listen to each one. And once all the explanations are out of the way, talk about the process and what can be done to improve it. It may seem like overkill, but determining next steps and documenting the actions that are to come out of the mediation session will provide clarity, which was likely the problem to begin with.

Respect

Mutual respect is something I preach. Yes, there will be a hierarchy within any organization, but that doesn't mean someone's ideas are more valid just because of their position. Respect is a two-way street, and it is something that must be earned over time. Communication goes a long way regarding the respect and trust that creative teams require to operate smoothly.

Speaking from experience, there are a lot of egos in the field of technology. It can be challenging to navigate these personalities, but understanding that everyone has something to offer will lead to open-mindedness. I never want to be the smartest person in a room. I want to have something to offer, but I am also there to learn from others. Remember that each person you work with is a human. They have feelings, aspirations, and insecurities, just like you. Bringing kindness into the dynamic helps me connect with my team and earn their trust and respect.

Healthy Noes

The word "no" is powerful. It'll save you and your team a lot of trouble when exercised properly, but I try not to say no without a why. In a creative space, it can be vulnerable to share ideas, which means that if someone is coming to you offering their thoughts, it may have taken courage to do so. Give each of these moments a true listen. Don't zone out, don't assume it isn't going to work out, but listen actively and then compose your thoughts.

Extended reality production is an extremely collaborative process with so much still unknown that you want to encourage members of your teams to come up with new concepts and share their ideas. You never know when one is going to be a winner!

If the answer is no, that is completely fine and reasonable. There may be a multitude of reasons why you can't or shouldn't do something, such as budgetary constraints, time limitations, or simply that it wouldn't work with your specific project. Explaining why you've come to this conclusion will encourage strong communication and collaboration, as well as instill in your team that ideas are valid; they just may not be a fit.

From scheduling to task management, and even the soft skills too, production's role in an extended reality application touches every part of the process. There is no way to craft a cohesive product without it. Lay your foundational tasks and schedule with care and consideration. Do a gut check every so often to make sure priorities haven't shifted. And be the champion of a healthy team dynamic because teams that communicate and collaborate put out better designed extended reality experiences.

8

Environment

ENVIRONMENT DESIGN IS a great path for fine artists and digital artists. And while I don't see many entering the field just yet, architects, interior designers, and landscape designers already have plenty of transferable skills to bring to extended reality production. Designing environments is more than just digitally modeling 3D objects. You must think about the whole space, every little detail of it. When working out the preliminary concepts of environment design, make sure that your choices have intention and that they will make sense in an immersive world. Whatever you design will be produced by a 3D artist or team, and then any interactivity is programmed to function for the user. While most of this will be applicable to virtual reality production, the same consideration with style, modeling, and other elements that make up the visuals of an experience can apply to augmented reality.

A chicken-and-egg situation for me is spatial planning and art style. If these components of a project are being handled by separate resources, they can occur simultaneously, although at some point these elements of design will need to merge together. These two environmental components are independent of each other but can also inspire one another. If handled by the same resource, it may be more powerful to start with art style first, as creating concept art can open your mind to what the space is going to look like generally.

Art Style

The exciting thing about creating extended reality spaces is that there is so much yet to be defined. This especially comes into play when we talk about art style. I've noticed that the majority of experiences go down one of three paths. The first is realism, or at least as close to realism as the hardware will allow. The second

95

is futurism, which uses a lot of black and neon and looks like something I could classify as "sci-fi abstract." The other is low poly, which has a cubic look and is highly optimized for the tech limitations of the hardware today. While other examples exist, most experiences seem to fall into these buckets. The great news is that there is an opportunity to define your own style and bring a new look into the extended reality industry.

Think of how you plan to incorporate form, color, and composition into this experience. What feeling do you want to draw out with these choices? When working through the art style for our gardening game, Loam, we knew we wanted a mature color palette that blended in with nature. Due to the large number of objects and interactivity we were trying to achieve, the form of our objects needed to be low poly, meaning simpler shapes that are easier for the device to process. And when thinking of the composition, we wanted small plots of land surrounded by mountains and a valley so that the interactive area would be constrained but not feel small.

The art style must serve multiple purposes—aesthetics and technical function. Unlike a painting, where once the paint is on the canvas you can simply see it, you must imagine that composition in your mind and then make sure it can actually run on the hardware once created. It's completely doable; there are just some extra steps and considerations to take to get you there. Something else that people don't realize is that colors, text, and sometimes objects themselves must be brighter and larger to look correct in virtual reality. This can be tough if you're designing these mockups and concept drawings on a computer screen, as things might look a little too bright or over the top before viewing them in virtual reality. That's why it's so important to check your work as you go, even if it's still conceptual.

Spatial Planning

When working through your spatial layout, consider the floorplan, key object placement, and flow of the space in a 3D environment. As you determine the layout of the space, remember you get to leverage full 3D and all of its depth and dimension. While this is important when blocking out the location of objects in a room, use it to your advantage for surprise and interest. Leverage walls, windows, and doors to make a space seem more expansive than it really is. Always thinking back to optimization; you can get a lot of efficiency by using visual tricks, such as flat elements behind windows or off in the distance.

The flow of the space, what you're going to include, and what you'll limit should guide placement of objects. Don't place objects of interest just out of reach, especially if they're supposed to be interactive. Consider the playspace, or area in which the user will physically stand and walk around. If you anticipate users to max out at an eight-by-eight-foot space, try to group your interactive objects together in clusters so that they can use the experience without having to teleport so often. Sketch out your locomotion boundaries as you create the space. Identify if you're limiting their movement to specific areas or if you will allow them to move freely throughout the environment.

A tool that I like to use for both spatial planning and art concepting is Open Brush. Built upon the foundation of Tilt Brush, an incredible virtual reality art program by Google that was later made open-source, Open Brush supports file exporting. You would likely never use these files in your final product; however, it's a great starting point if you've sketched something out spatially and then need to match up the positioning and scale in a 3D art program. Alternatively, ShapesXR is a more collaborative tool, and it even lets you create "Stages," which can act as slides in a storyboard to show progression in your experience. It doesn't have as many artistic features as Open Brush, but it is fantastic for spatial planning and reviews.

Once a space is laid out, the details emerge. Bring in your 3D artists, share the vision of the art style and spatial planning stage, and get to work creating something beautiful. It's quite possible you included your 3D artists, or a lead artist, as a part of that planning stage—I'd recommend it. However, if working with an outside team, share any of these materials so they understand the vision and composition before getting into the detailed modeling. If your project is going to leverage assets from another department—such as engineering—work with them to get the appropriate file type for optimization. Most extended reality experiences will leverage FBX, OBJ, or glTF files, so if they can provide one of those exports, even better. I'm not an artist myself, but after observing my team for years, I've collected these highlights to get you started.

3D Modeling

Whether or not you start from preexisting engineering files or are creating something completely new, leverage 3D modeling software that can support export for extended reality preferred file formats. Blender has become a leading choice for many 3D artists, not only because it's free to use but also because of its robust and ever-expanding features. In addition to your 3D modeling software, consider tools such as Photoshop to modify textures and even Unity to make simple modifications to models.

Most 3D models start from simple shapes such as rectangles, cubes, and triangles. Start blocking out your models in a basic way, and then add more detail as you go. Consider the size of the models as you work on producing them, as their scale will be directly implemented into the experience—whether virtual or augmented reality. In most cases, these models should appear life-size, so start with the proper dimensions (or as close as you can get) as you build them out. If you're working directly from physical world specifications, there won't be any wiggle room for adjustments. If you are implementing objects that are less precise, view them in your platform as soon as you can. You may notice that the scale is off, or objects feel too small. This is a common feeling when creating interactive objects, as the precision of interacting is still a little clunky. Feel free to scale objects slightly larger than you'd expect to make their interactions feel better in a digital deployment.

Optimization

Speak to your production team about the hardware these models will be running on to identify any limitations required. This may indicate the need to reduce poly count, although that doesn't mean you have to go full "low poly" art style. You can reduce the number of surfaces of your objects (especially the background objects) and judge visually where that cutoff between looking realistic and too cartoonish is. It can also be helpful to leverage texture atlases, which combine many texture files into one larger map. Think of a large image that contains a grid of smaller images, and each image is mapped to a 3D surface of one of your models. The program is referencing the texture atlas to display the correct texture on your model.

If you're deploying on a computer-based system, you can pump up the graphics quality in your production process and create near-photorealistic models and environments. This is especially magnificent when pairing those graphics with the optics of a high-quality headset, such as Varjo devices. Higher-fidelity graphics will become more common as advancements in processing power and optics reach smaller, stand-alone mobile devices. Something I like to look forward to is the fact that whatever models you produce today will only look better down the road as new devices come to the market!

Other factors make up the entirety of environment design, not just 3D models. This can include animation and movement, visual effects, lighting, and audio. Without these other elements, the environment will feel stiff and lifeless. You don't even need characters in a space to bring it to life—it can happen with other ambient features. Adding these environmental factors will contribute to the feeling of presence we're always striving for in an immersive experience. Don't treat these elements as an afterthought; think about them, and evolve with them as the production process progresses. While you need to get your base environment complete early on, these other factors can be incorporated and adjusted as you go to ensure the experience feels right. The experiential subjectiveness of that feeling may warrant tweaks toward the end to finalize the product.

Animation

Animation and movement add something I once heard referred to as "juicy" in a digital environment. I've encountered the term "juicy" mostly in the video game production world, but it has a place in extended reality production as well. As best as I can describe it, it's the little stuff, the minor details, that add anticipation and a feeling of satisfaction to what you experience around you. Think of the little bounce you get when you tap on something or a tiny firework when you collect an object in a game.

You may not notice if these elements aren't there, but you have a more positive experience when they are present. Bringing these little elements of movement into extended reality takes thought and careful implementation, as too many contribute to a cluttered and confusing experience.

In addition to the little details, larger expected animations are almost a requirement. If you have a machine in an environment, the belts should move, the levers should go up and down, and the gears should turn. There is an expectation of movement in the physical world, so bring that expectation into the digital world. Animation can apply to objects as well as characters. Production teams typically use something called "timelines" to string together a series of animations if a sequence needs to occur in order as part of a storyline. If an object or character is going between two or more states or poses (especially facial poses), an animator may use "blend shapes" as a more efficient way to visually pass between those states. This method can be easier as well as more optimized than using full animations.

Effects

Just like other digital media, there are a plethora of effects that can be added onto art assets to bring life and interest into the environment. Particle effects are commonly implemented to represent moving, nonsolid elements such as fire, smoke, water, and dust. Building onto those physical world elemental examples, particle effects can also be used to indicate the success of an action—think about little stars bursting around an item you've just successfully interacted with. When implementing a particle effect, you'll have the opportunity to define the point of origin, how much and how quickly the particles disperse, and the look and size of the particle itself.

Shaders are another effect that can add either realism or a stylized look to an entire digital environment. In addition to a stylistic change, shaders can also add motion to a digital asset. For one project we produced, all of the assets were modeled in 3D, but we applied a shader to make all of the elements appear as if they were 2D so you felt like you were inside a comic book. If you decide to use shaders and particle effects in your virtual reality experience, do some benchmark testing for their impact on frame rate to confirm if it's okay to incorporate these elements. While they have the potential to make a big visual impact, they can also bring performance down, which could cause an uncomfortable experience for the user. This is all highly dependent on the optimization of your assets as well as how interactive your experience is, so there may be hope for you to use these visual effects for added wow factor.

Lighting

Lighting is another touchy subject when it comes to extended reality, specifically real-time lighting. In an augmented reality experience, ensuring your models have lighting by way of a source of digital light to create shadows will make them look more realistic. Test your model placement in a variety of settings in the physical world—indoors, outdoors, bright, dim—to ensure that the lighting settings you've chosen look real enough when placed in the physical world. Because in AR you're not going to be rendering entire environments, real-time lighting can work in most applications.

In virtual reality, unless using a powerful system, we try to bake lighting when possible. Baked lighting is a static form of lighting that has been embedded into the 3D models, which is more optimized as the lighting has been rendered onto the object once.

This information is stored in a texture called a lightmap. The lighting will appear the same on the model, every time. Using real-time lighting is acceptable within reason if your system and experience can handle it (meaning it does not affect the frame rate significantly, just like other effects). If your experience can't use real-time lighting, there are other ways to trick the eye, such as the use of shaders and generated shadows. In our gardening game, we wanted the time of day to change but couldn't use real-time lighting, so we added shaders to procedurally change the colors of the objects throughout the day. All objects in the environment would have the brightest colors in the morning, most vibrant around early afternoon, orange and pink hues in the evening, and a bluish dark tint at night. When you're in the experience you see the sun and moon moving throughout the sky, but the lighting changes are a complete trick, and I'm not ashamed to admit it.

Audio

We've already touched on audio design, as well as sourcing and quality, but it needs to be mentioned in environmental composition as well. As all the visual elements come together, motion is added, and objects are designed, keep audio and its impact on the experience in mind. When you bring the environment into your development engine for the first time, start to place sound effects or ambient audio in the scene so that you can feel the balance and course-correct if something is missing or off.

As a production team, we must occasionally make assumptions for what something sounds like, especially when making training content that's outside our area of expertise. I recently had a project where we got the pipe bursting sound wrong, but how would we know that if we've never witnessed that on a job site? Let your subject matter experts, or someone who is close to the material, examine the effectiveness and realism of the audio in the middle of your production process (not just at the end). If the audio is more abstract or creative, still get a second opinion from a team-mate with a good ear. While it's important to stick to the original design, some things you can't predict or finalize until you pair the audio with the visuals and experience it.

Characters

Characters can bring life into an experience, especially one with narrative. They can also make a single-player experience feel less lonely. We leverage characters quite often in training simulations so that trainees can have a guide or instructor in the experience with them. Characters can also demonstrate movements or tasks, which can help users understand how to perform interactions in virtual reality. Implementing characters may not be a fit or need for every experience, but there are some considerations if you're incorporating them into your application.

Character Models

There are great tools out there for creating character models for an extended reality experience. Just like your typical 3D models, characters must also be optimized for the hardware you're using. Many character creator tools have settings to export these assets in different fidelities so that you can choose the best fit for your project. The tool my team uses is called Character Creator by Reallusion; however, there are many systems to choose from. We try to have a cross-section of representation when we design our characters, and I encourage you to do so as well to bring diversity and inclusion into extended reality.

Many of these systems will also include libraries of animations that you can leverage as the basis for character movement. If you require specialty animations, motion capture could be a great path to producing the animation files you need. For years, we used an inertial body sensor motion capture suit. It was only recently that we started using the Move.ai system, which combines artificial intelligence and computer vision to compose motion capture data from multiple camera feeds. This means we can do motion capture sessions with anyone, anywhere, no suit or hardware required. There is still a good bit of cleanup for our animator to do, but it's made a huge impact on our production workflow.

Uncanny Valley

There is an odd occurrence that happens in virtual reality called the uncanny valley—yet another feeling to add to the list of subjective checks. The uncanny valley occurs primarily with characters that are meant to be human but aren't quite there. I have found that in social experiences where I'm represented by an avatar, the less realistic and more cartoony systems actually make me feel more comfortable. Having a stylized avatar, such as those of Meta, AltspaceVR (RIP), and RecRoom, make me feel less strange. Seeing my friends in these forms is more pleasant than something that looks almost like them but not quite.

The best way to prevent this with your character models is to ensure they have more emotional movements in their faces. We use a combination of eye movement, facial puppeting, and mouth movement to bring a face to life in extended reality. If you're not going heavily into a stylized character look, ensuring that your characters or avatars have proper human proportions will also help prevent the uncanny valley effect.

As you compose your environment, remember this digital space is more than just an arrangement of 3D models. At the risk of sounding like a broken record, develop these assets with optimization in mind from the start. Art is going to have the biggest impact on the performance and frame rate of an experience. Movement and sound have enormous influence on how immersive an environment can be. Think about the physical movement of the user as you work through spatial planning and the flow of the world you're creating. You're building something incredible, and now it's time to sync up with the developers to bring in the user flow and add some interactivity.

9

Prototype

WHETHER YOU CALL it a prototype, phase I, or proof of concept (POC), this phase is a necessary step in testing things out, evaluating what works and what doesn't, and shaping the pieces of a long-term plan. Many people refer to this as prototyping; however, I like to save that word for the experimentation and iterative development that goes into creating the pieces of an interactive experience. When referring to the phase of work itself, I suggest that every extended reality project either goes with a proof of concept or phase I to start. I define these differently as various projects have timeline requirements or specific goals that contribute to this categorization.

Proof of Concept

I classify this as the smallest scope, fastest timeline, and most budget-conscious production option. Based on the experiences I've produced, a proof of concept never goes beyond the use of getting more buy-in or testing out whether extended reality is the right solution for the organizational goals. You may be able to take components or inspiration from the project files, but the code itself is created in a way that is quick and dirty, meaning that it needs to be functional but does not need to be foundational. It's difficult to build additional features or expand on something that's been developed in this manner, which is why I am very clear about the intent of developing a POC-level experience.

A proof of concept can still be a great first step, as it often takes someone experiencing the technology to understand the impact and power of it. Because of the limited amount of content on the market today, it can take a customized POC to get the point across to leadership for additional budget and investment into the technology. If you decide that a proof of concept is the right first step for your organization, design a simplified version of your user flow. Consider your audience as well when modifying your design. If the overall goal is a series of training experiences in manufacturing but you must adjust it for executive leadership, ensure the steps are clear as these individuals likely don't have the same type of experience as your facility workers. Craft your proof of concept to show them the power of the technology applied to their industry but without all the nuanced complexities of the training content. This will allow them to get the big picture while keeping scope limited.

Phase I

I consider phase I as the initial project in a (hopefully) long series of projects, or content development work. This is the best path to take for long-term expansion, so if you or your team already know you are going to deploy something functional, start with phase I instead of a proof of concept. When embarking on this development phase, have as much foresight as possible to know how you'll expand the content. Is each experience you produce going to be deployed as its own application? Will you need to make modifications to the content over the following months or years after it's released? Answering these questions early will help you put the foundational pieces in place and allow for more optimized scaling when it's time to do so.

When you think about what content is going to be best for phase I, consider the reach of your application. Is there a particular product or product line that is a best seller? If so, that may be a great fit for phase I of an augmented reality visualization tool. You want this to be wide-reaching so that you can get feedback and measure the success of the development effort. One of my manufacturing virtual reality clients chose a difficult-to-train-for machine as their phase I content, as they knew they'd be able to get the most impact to the business with this choice. By proving they were able to reduce training time and increase knowledge retention, they were able to continue building out additional phases because of the success of phase I. Choose your content with care, and then identify ways you will measure its success so that you'll earn the opportunity to implement more extended reality solutions in the future.

Prototyping various features in an immersive technology solution is an iterative process. You can design with as much detail as possible on paper, but these features will always need some tweaking once implemented and experienced on the hardware itself. It's necessary to iterate, and it's healthy to admit when something is not working. The more time that is wasted trying to fix your original design, the less time you'll have to come up with a new solution. I often must pull my developers outside of their own heads when they are staring at an issue too long because they can only see one path to fixing it and are trying to brute-force something into working. Rely on your teammates to help you break out of your mental blocks in these scenarios. Talk about it, even with members of a different department. While I'm not a developer, I've had developers on my team say that I've helped them think of a new solution, and it's worked. I use the fact that I'm not a developer to my advantage as I don't know all the ways something should technically work. Because of this, I can make out-of-the-box suggestions to break those mental barriers. The three elements we find the most difficult in the prototyping process are interactions, locomotion, and tutorials.

Interactions

Interactions are the toughest part of making an extended reality experience feel real. They have the potential to add high levels of immersion or introduce frustration to the user. While there are some standards in interactivity, such as using the grip button on the controller to grab items, there is plenty of room for experimental and new interactions to be designed. In our gardening game, we wanted people to be able to harvest crops and pick flowers but not do so accidentally as it takes days in the game to regrow them. We couldn't just use a simple "grab" interaction as that could add to frustration if players spent all this time curating and tending to their garden only to accidentally wipe out a crop. We decided to implement a pulling motion in combination with grab to make them feel like they were pulling the crop off the plant. It's a new action to learn, but once you realize it's a pulling motion (like reeling in a fish), it makes sense and allows the action of harvesting to be more intentional.

Think about what action you're trying to simulate. How is it done in the physical world, and how can that be translated into an immersive experience with either controllers or hand tracking? In my actual garden, it's not so hard to pick a tomato off a vine, but I also don't go up and grab near my tomato plant unless I'm ready to pick it. In the environment design chapter, we discussed how you must make text and objects larger than life. The same can be said for motions. The level of precision you would have in the physical world, especially with dexterity in your fingers, doesn't translate unless you have additional haptics (like gloves) to help with that precision. In one of our training simulations, the trainee was supposed to put a tag on a pipe to signify there was an issue in their work area. While this seems like a simple grab and place action, we needed to widen the collision area, or place at which these two objects intersect, to make it feel more natural in the experience. We weren't teaching them precisely how to place the tag but that the tag placement was a required step in their process. The goal of the training experience allowed us to make this change. Reducing the precision helped with the user experience and contributed to a more seamless interaction.

Locomotion

Locomotion best practices are fairly established, especially regarding comfort for the user. We have two main methods—teleportation and smooth locomotion. However, it's important to prototype how your user will move throughout the environment. If free teleport is the option you choose, know that the user will be able to pop around the entirety of your space. You may want to set boundaries, and if you do so, make sure there is some sort of visual indicator in the scene that signifies an area is off limits. We often use decals and symbols on the floor to identify the places our users can and cannot go. In testing, we sometimes find that free teleport is a little too free—especially for newer virtual reality users. If the experience warrants it, we'll leverage something we call teleport nodes or hotspots to draw users to points of interest in the room. These can be used as the only guide to teleport, removing free teleport altogether. Or they can be used in addition to free teleport to give the users a definitive location to snap to if they get lost, confused, or overwhelmed.

The placement of these hotspots is something you'll want to prototype and iterate as you go. You'll want to ensure there are enough interactive items within reach of these hotspots and that the hotspots are close, but not too close, to these objects. If you get teleported too close to something, you can feel unsettled and even end up inside of a 3D object or wall if you're not careful.

This is another example of something that can't be finalized until you experience it firsthand.

If you want to create a new method of locomotion, by all means give it a go! I've loved seeing games leverage arm movements or other creative approaches to help the player progress around their area. I recently enjoyed zooming around a zero-gravity virtual environment using mini jet packs on my arms—until I didn't. The highest consideration when designing a new form of locomotion is comfort. Prototype your ideas, and try to test these out with people who may not have as strong a stomach as you. There are certain people on my team (myself included) who are designated to tell the development team whether something will make people sick or not. My lead developer is one of those people too, and together, we help ensure that our users won't feel nauseated by taking that burden on ourselves. It's not a fun task, but better for us than our end user, and we always keep our fridge stocked with ginger ale.

Tutorials

Tutorials may not seem like something you have to prototype and iterate on—just tell someone how to do something—simple as that, right? Wrong! Tutorials can make or break an experience in the first couple of minutes, and explaining to people how to perform actions on a technology they may never have experienced before can be a challenge. Ideally, this will not be the circumstance forever, but for most of our business use cases, we must assume that our audience has never used extended reality in their life. We not only have to teach them how to use our experience, but we must also introduce them to immersive technology in general and do it concisely.

While the full and complete tutorial may not be implemented until closer to the end of your development cycle, try to incorporate pieces of it as soon as you are able. As you'll be teaching people how to use the technology in general, include controls mapping and how to select items and objects first, and then go into specialty interactions. If they open the experience and there is some sort of login screen or menu, don't forget to show them how to use that menu and what controls to use. Include animations or examples of each action when you can. It's great to pull mini examples of interactivity out of your experience to include in your tutorial for familiarity as the user progresses. Test this with other members of your team, and observe their reactions. Confirm if the instructions were clear or if you need to add more detail to get the point across. Just because something makes sense to you, doesn't mean it's understandable enough to others, as you're likely very familiar with how things *should* work. While you want to be concise, you need to ensure you're not leaving users confused, so prototype your tutorial with the least experienced user in mind.

When you're in the middle of producing a proof of concept or phase I, it can be helpful to have parameters to limit yourself in the beginning while you work out all the features and functionality. Something that's common in video game production is called a "vertical slice," which can be applied to extended reality as well. A vertical slice is a complete loop (meaning it can be played or experienced with a start and completion state) and includes all your essential features, framework, and systems. The limiting factor here is that it does not include all your content. Choose one preferably simpler piece of your content to leverage in this vertical slice, and incorporate your features into one complete playable system.

Systems

Bringing together your systems in a vertical slice is an integral part of creating a product that can be experienced. You've designed and prototyped all your components separately, and the incorporation of these elements will produce something that is usable. Systems include any elements that would need to be combined to produce this playable experience. Building on the framework and engine in which you're developing (think Unity or Unreal), there are additional layers that are required to run your experience on the hardware, which can include OpenXR, Snapdragon Spaces, ARKit, and more. Each type of interactivity that you hope to incorporate is likely considered its own system. Even the method of locomotion you choose is a system that needs to be integrated.

In addition to these technical systems, you'll need to incorporate the systems that allow you to show linear progression, whether it's a timeline that hosts all of your animation and audio timing information or something else. If you have a need to manage scoring or calculations, incorporate that as well. Depending on what the experience is, there may be choice-based outcomes, which must be applied to generate a complete application. Everything that is required for the experience to run (and be completed, most importantly) needs to be merged and verified that it works together. Notice I said "required," so if there are "nice to have" features or interactions, you do not need to include them at this time.

Content

While all your systems need to be incorporated in a vertical slice, the exact opposite is true about content. This is what you want to limit as you test out the full experience loop. Similar to how we identified the content that would make up phase I, think even smaller than that. If you want to show how a machine works in virtual reality, can you focus your vertical slice content on the start-up procedures of that machine? If you're creating an augmented reality product visualization tool, can you implement all placement and interactivity with one product? The point of the vertical slice is to create a polished and smooth user experience with limited content.

If you are producing a storytelling or narrative experience, narrowing down the content may be more difficult. If this is the case, try to focus on the intro or opening of the story. Set the scene and leave it on a cliff-hanger—television shows do this, so why not your vertical slice? Make sure as you think about the narrative content for this limited version of your experience that you incorporate all the base functionality and interactivity into this shortened piece. As the point of the vertical slice is to test all necessary functionality and how it fits together, you don't want to design content that won't support all these requirements. Ideally if this is part of phase I, you will build on this vertical slice, not scrap it, so design the content with that in mind to reduce rework after the project expands.

This phase of the project should start to bring life to your designs and environmental assets. It's a great time to work out some of the kinks before preliminary experiential user testing. I keep teetering on the line of "make it good" and "don't finalize anything" for a reason. An extended reality experience is in my opinion one of the most difficult media to produce. You are bringing together so many elements that in other industries would stand by themselves. Yes, the film and television industry leverages audio and visuals to take their audience on a journey. Yes, the gaming industry adds interactivity on top of that to bring you into the story as your own character. But with extended reality, you can feel like you're living it and it merges the experience with your reality like no other platform can.

It's too early to know how all the elements will fit together in the end, and user feedback is a crucial step in getting to a final product. When you prototype all these components, be cognizant that all of this is still evolving. Make it good, but don't finalize anything. You will be so close to the product, it's easy to think something is simple and useable when in reality, you think that because you're the one who designed it. As we move into testing, keep this in mind and don't be afraid to go back to the drawing board. Hold on to this mindset of innovative thought as you progress through the next stage of production.

10

Testing

Obtaining feedback outside of your production team will be crucial throughout the entire development process. While some may say it's never too early to start testing, I'd change that to say, "It's never too early to get feedback," which can happen at any time throughout design and preproduction. There needs to be some meat on the bones of your experience before testing makes sense in virtual or augmented reality. You can get in front of a test audience as soon as you have a playable loop completed, whether that is a proof of concept or the vertical slice of your phase I.

Try to present something to a sample group of your end users or project stakeholders as soon as possible. This can sometimes be tricky because you need to find people who understand when things aren't complete. These individuals need to come in with an open mind but also be critical of what's working and what isn't. If you're focusing on training content, precision is key. If you're producing something more narrative, subjection comes into the mix. This chapter will focus on more subjective and experiential testing as well as the overall flow of actions. Chapter 13, "Quality," will focus on alignment with your project requirements, design documentation, and technical functionality of the program. You'll want to test the tutorials, interactions, spatial placement, and flow of the experience, at minimum.

Tutorials

Ideally, you will already have at least a limited tutorial in any experience that makes it to a test audience. Observe whether these tutorials are helpful and make sense to the user. Gauge the experience level of your testers, and note the differences in interpretation of your tutorials for a less experienced extended reality user versus someone who has leveraged the technology before.

If the individual is less familiar with the technology, are there any points throughout the tutorial where they are struggling? Document whether it's the instruction to learn the interaction that is the challenge or whether it's the interaction itself. If it's the instructions, identify what isn't clear and discuss what would make sense to that user so that you can improve the tutorial in the iteration stage.

Also account for the time it takes for the test user to complete the tutorial. While you want the tutorial to thoroughly hit on all the appropriate topics, it needs to be completed fast enough that it doesn't become a burden on the users. If the individual is already familiar with extended reality interactions or has already gone through the experience before, ensure they are able to easily skip the tutorial and that the option to do so is clear.

Interactions

Testing for interactions will vary depending on what functionality and controls you have set up in your experience. While the initial understanding of these interactions will be tested in your tutorial, you also want to confirm that as the interactions come up in the experience, they make sense. Does the instruction the user learned in the tutorial translate over to the real-time application of that knowledge?

It's important to keep all users in mind, but the most likely frustration with interactions will come from a first-time user or someone who is unfamiliar with extended reality. Because interactions in extended reality are spatial, meaning you're moving your arms around quite a bit and grabbing digital objects, there are some assumptions that come along with that. People already have an idea of how to perform these actions in the physical world, so as they go through your experience you may see them try to do something similar to their physical world actions in the virtual world. If those assumptions don't match up with how you've set up an interaction, make a note of that. It may be impossible to set up the interaction identically to how it's done in the physical world, but if there is an incorrect assumed action that test users continue to take, a change in the interaction design to mimic that may be worth consideration.

Spatial Placement

Testing for spatial placement, or the placement of objects and layout of the virtual assets, is something that is important to test for in both virtual and augmented reality experiences. You want to discuss with your virtual reality test users whether the placement of the objects in their space makes sense to them. Ensure that all interactive items are within reach as they progress throughout the experience. If something is out of reach, determine whether it is an accessibility or height issue, or whether they're limited by the boundaries of their playspace setup. If the user needs to move throughout the space via a method of locomotion, are they able to do so? Confirm that they can easily get from room to room or around the environment they are currently in without challenge.

Spatial placement can also relate to the placement of augmented reality assets in the physical world. Have a discussion with your testers about the scale of the objects they are placing in the physical world. Do they look correct? Identify with your test users whether objects appear to be floating or whether the lighting is drastically off from what they'd expect in an overlaid asset. If the experience allows movement of the digital objects, note how that movement around their physical space functions. Observe the test user and how they transform the digital assets in the physical space, and note any issues.

User Flow

The user flow, or flow of the experience and progression of the content, will likely have some subjective feedback during this process. If the test user can complete the full loop successfully, great! That is the goal here. If they aren't able to complete the full user flow as you have it implemented at this time, you will not be able to complete your test session. As they go through the flow of the experience, observe their choices. Ask them to talk through their thought process out loud so you can understand what they're thinking as they encounter each point of potential divergence within your experience.

Similar to the other test categories, we want to determine any points of confusion or frustration for the user. If there is a point at which the user gets stuck, document whether it's because the options were not clear or whether they did not feel there was an option to choose from at all. In the development process, it's important to incorporate fail safes for these potential stopping points so that the user doesn't feel like they're locked in the experience. If you do not yet have these fail-safes in place for this test session, you may need to provide guidance to help the user progress through the flow. Don't be too quick to jump in and help them so that you can document the issue. However, it is okay to assist them in their progression so that you can complete the test session.

Try to have an open mind when you're going through these test sessions. Ensure that you're listening to the tester, and do not get defensive. You want to encourage honest feedback and a healthy dialogue, and if you get defensive, they may not feel comfortable sharing their opinions. It can be difficult to hear someone express that they don't understand something you think works perfectly. Swallow your pride and take a breath because it'll make for a better product in the long run. Don't tell them how something is supposed to work before letting them explain what they're experiencing, or you may miss out on valuable information.

Test documentation is something you'll need to prepare before performing these test sessions. Assembling this documentation will help you stay organized, focus on the information you need to gather, and encourage consistency across all of your test sessions. While there will be nuances and variation across your test users, building off a consistent plan will allow you to see patterns and commonalities across their user experiences. If you're having difficulty prioritizing what feedback to act on, these patterns will help inform those decisions.

Testing Brief

A testing brief is a document or presentation we create and either review or provide to the test users to give the experience context. Details may include the goals of the experience, content parameters, and an overview of expectations when it comes to reporting their feedback. While it's good to provide relevant information, keep the overview light so that the information doesn't affect their test experience. You want to capture their initial reactions to the experience, so telling them too much may spoil some of the surprises or lead them into one reaction or another. If you're testing with a proof of concept or vertical slice, provide enough details to make up for any gaps in the experience that currently exist.

In addition to contextual information, include specifics about what type of feedback you're looking for. If you're testing for spatial placement, environment, and art, mention that, as they can focus their thoughts and observations on that. When testing the overall experience, make sure to ask for comments on interactivity and flow. While you have this captive audience, use it as an opportunity to talk about presence and immersion. The only way you'll be able to test for these elements is through user testing, so consider including definitions in your brief.

Intake Form

If you're performing the test session with the user in person or virtually, you'll want to create an intake form that you can consistently use across testers. This should include the key items you'd like your moderator to observe as well as a place for comments under each point. In addition, you can provide a scale from negative to positive that goes along with each item to quickly gauge the user's reaction and use that to compare data across all testers. At the end of the form, there should be a space for the moderator to take additional notes that are specific to that user's session, including suggestions from the test user. While each project's needs are different, start with a baseline of tutorial, interactions, spatial placement, and user flow. Expand from there to include any specific points you'd like to incorporate for your sessions.

If the user is testing on their own, it may be more beneficial to provide a post-experience questionnaire. This can be created in any form builder platform or sent out as an editable document. As you craft the questions, read them over to confirm there are no leading questions, or questions in which you prompt the user into thinking one way or another with your phrasing. For example, instead of asking, "How much did you enjoy the experience?"

consider, "Was the experience enjoyable? Please explain." The first phrasing leads the user into providing a positive answer, whereas the second option lets them choose if they respond with a positive or a negative. You can create the questionnaire specifically for your extended reality application and the goals of your testing session. Feel free to include questions about individual interactions in addition to the overall experience.

Session Recordings

While you may not always be able to record each session, do it whenever possible. Always make sure you have consent to record before starting, and save your recordings in a consistent location with a common naming scheme that your team can access. When subjective feedback is at play, which happens a lot with extended reality production, it's nice to refer to the test users' exact words and actions to help explain their reactions to the development team. Video when testing virtual reality interactions is extremely helpful both from outside of the headset as well as the user's first-person point-of-view footage. By seeing the recording of their body and movements, we can identify how they performed each action physically. With the first-person point-of-view footage, we can match up how that action occurred in the program and see how they were interacting spatially.

When an interaction is difficult for a user, look at the first action they performed when attempting that interaction. It's possible that their assumed action could lead to a better, more obvious way to design that interaction. That won't always be the case, but if you notice a pattern across testers, this assumed action could alleviate a problem area. It's difficult to determine that without video to look back on. In addition to interactivity footage, take video or screenshots of the experience when something isn't functional or looks off. In one of our projects, one of the characters would start walking on the ceiling, defying the law of gravity. It was startling, and we wouldn't have been able to understand or replicate what happened unless we had that footage.

Performance

While we won't focus on technical quality and bugs for a few chapters, it's never too early to note your application's performance metrics. Identifying issues in this preliminary stage will help you reduce rework down the line. By documenting these metrics early, you will also have benchmarks to test against as you add more features, functionality, and effects. Test these metrics with every new build so that you can isolate the elements that have a negative impact to only the changes made in that release.

Metrics Tools

There are a variety of metrics tools that developers use, and they can vary by hardware. One of the common tools we leverage is called OVR Metrics Tool (originally Oculus VR, now Meta). This tool can be turned on inside the headset and provides an overlay that shows a variety of useful information, such as frame rate, CPU usage, GPU usage, battery percentage, and memory usage. This isn't something you'd want to leave on all the time as it's directly in your field of view, but it is extremely helpful in seeing real-time spikes and dips in your metrics, which allows you to pinpoint the issue.

While OVR is specific to Meta devices, there are other similar tools and even services that can help provide this information.

For example, PICO has their own Metrics HUD, which can be enabled when devices are set to developer mode. Whichever method you choose, make sure to be consistent with your logging so you can identify problems before they become a larger issue. You would likely only use metrics tools with your own internal production team; however, if you have what I like to call "super testers" as part of your test user group, they may find value in learning how to use these tools as well.

Frames per Second (FPS)

Frame rate is calculated as the number of frames the device's GPU can send to the display per second, noted as frames per second (FPS). Frame rate has the highest potential to render an experience unusable. That may seem like an exaggeration, but a low frame rate will cause choppiness for the user and can lead to nausea and headaches very quickly. The benchmark for what is considered a good frame rate continues to increase, which is exciting for the advancements of technology and what's possible. Good is considered anything over 60 FPS, and great is over 72 FPS, with newer standards reaching 90 FPS and higher. From personal experience, I don't get sick until I spend a good bit of time in something that's in the low 50s.

Because of the enormous impact frame rate will have on your user experience, immediately note when you identify a dip. There are three factors that typically cause this. The first is the art assets themselves, which is why it's so important to optimize, optimize, optimize. If you notice a dip when looking at specific objects, you may not need to optimize the entire environment, just those assets. The second factor is effects, such as shaders and particle effects. While they are a great way to add visual intrigue and excitement to your virtual world, you may need to rethink how they're implemented if they are causing a problem. And finally, the interactions themselves can cause frame rate issues. This is less likely to occur than the visual assets, but if you're packing too much into an experience, it can have an impact.

Something that has nothing to do with production but can also have an impact on the device performance is battery level, so if you're noticing poor performance out of the blue, make sure you're not running too low.

CPU versus GPU

As you'll see CPU and GPU stats listed in your metrics tool, it's good to understand what they mean and how they work together. The CPU, or central processing unit, performs many computational calculations very quickly to execute programming. It's basically controlling the flow of all the operations a program needs to activate for an extended reality experience to work.

The GPU, or graphics processing unit, renders the graphics of an experience, which allows all the visuals to run. In addition to delivering graphics, the GPU can handle mathematical calculations as well. This means the GPU can execute scientific computations and work together with the CPU to deliver the experience to the user. When observing the numbers on our metrics tools, we typically see more spikes in the GPU than the CPU. This is because extended reality experiences are graphics heavy, requiring the rendering of full 3D environments at a rate of 60 FPS and higher.

As you go through this initial testing phase, keep your mind open to adjusting and tweaking the experience to be better for the user. Form your test user group with intention and be cautious of a "too many cooks in the kitchen" situation. Gather a good cross-section of your audience, stakeholders, and others, but don't overwhelm the process with too many people. When working through test sessions, remember to listen to the user, and no matter what, do not get defensive. By keeping this welcoming attitude, you'll gather valuable and actionable feedback that will only make your product better. The tough part now is paring it down and prioritizing so that you can iterate the design and fine-tune your deployable product.

11

Iteration

ONCE YOU AND your team have some preliminary test feedback, it's time to iterate. Please note that you should take anything your testers said with a grain of salt. You know your vision and your end goals better than they do and need to keep that in mind as you drive the next phase of development. However, don't let ego get in the way of valuable feedback from real players. This phase of iteration in the design of your work will go through cycles of testing and tweaking and more testing and more tweaking, until you have interactions, user flow, and spatial placement that suit the goals of your project.

If you find that something you thought was going to be the new standard of an interaction did not test well, it's time to reevaluate. Consider your audience as you work through these changes, and put yourself in their mindset. Go through the steps of asking yourself why something didn't work, how the test user thought it should work, and what you're going to do about it now. Try to get to the root of the problem before working on a design change. Once you have progress with your project and a new usable build, bring some of your key test users back into the conversation. Allow them to try your updated solutions so that you don't go too deep before confirming the new design will work.

Why

Determine why an interaction, the spatial arrangement of items, the user flow, or other components of your experience didn't work. Consider why an element didn't test well. Identify whether there were challenges because there weren't clear instructions or direction within the experience, or whether there is a problem with the system itself. Understanding why something didn't work will allow you to create an updated version that gets to the root of the problem.

While so much of the testing period is focused on 3D interactions and experiential behaviors, don't forget about your 2D user-interface elements. This can go for both augmented and virtual reality. If your users were having challenges with menus or activating certain tools and features that are dependent on UI elements, identify why this is happening. One of the most challenging components to design is an interactive 2D screen in a 3D environment. I've seen it in games such as RecRoom, and we tried to implement something like this in our gardening game as well. If your experience leverages a 2D screen (sometimes represented by a 3D tablet or object the user can hold and interact with), identify where users had issues and what the points of frustration were. Why did they have challenges, and where were the hiccups? Was it the overall use of the interface or the 3D object itself that delivered the interface? Those are two different design problems, rendering the why extremely important.

How

After you determine the key points of your experience you want to improve, it's helpful to look back on how your test users tried to perform these actions. As discussed in the testing chapter, this is why it can be incredibly helpful to have video footage of the test sessions. If you do have footage, review the movements of the user from both the outside perspective and the first-person point of view. Match up where they encounter a specific interaction or point of frustration, and then watch what they do with their body in the outside perspective footage. Likely the first movement they tried makes sense as either the instructions explained it or as an intuitive response to the situation in front of them.

If you don't have footage, you should still be able to look back on test notes or their post-session questionnaire to gather some insights into how they thought certain actions should work. Make sure to ask them this during your sessions as you encounter these points of frustration—most people will be more than willing to share their opinions on how they want something to work. If you notice a pattern between testers, that is an even greater indicator that their assumptions may be the best path forward when redesigning or iterating upon an action. Some of the best and most straightforward options will come from your testers, once again proving the importance of an outside perspective.

What

Now that you know why an element isn't working and you have an idea of how it could be improved, it's time to determine what you're going to do about it. What can be done to design a better interaction? Not all suggestions or options from your test users are going to be feasible within your timeline or with the resources you have to make these changes. You must balance outside feedback with your own production team's ideas and the limiting elements of the experience and hardware.

Take the time to write out, whiteboard, sketch, and play with the options that you come up with. It can be beneficial to isolate your top few choices and test these again to make a final decision on the best path forward. The changes do not need to be significant, either. Micro adjustments can be the difference in making an interaction feel right. My team is working on producing a creator tool for extended reality production, and we had a fun conversation recently about the delete object tool. Everyone loved the interaction itself, but the timing and visual indication of the object deletion wasn't as satisfying as the action.

We communicated about why it wasn't satisfying, realized that if the visual indication of delete occurred faster than what was designed, it would feel better to the user and that we could make a simple change that would take about five minutes. Try not to overthink or overdesign changes when working through the iteration process.

Updating your design documentation to reflect the changes is a crucial part of this phase of the project. If you don't update this documentation as you make changes, old design will be muddled with new design, developers may not be able to distinguish between the correct implementation directions, and mistakes will be made. This doesn't only affect the quality of the project but the timeline as well. Avoiding rework is one of the biggest drivers of keeping the schedule on time. Sometimes rework is unavoidable, but by making updates to your documentation and keeping these changes organized for your team, there will be a greater chance of successful implementation the first time around.

Annotation

Marking up your design documentation with updates and design changes is clearer when your team has a common method to annotate the existing documentation. Instead of directly editing the text or replacing reference images with new ones, keep the previous information for context. We commonly use strike-through text to indicate when something is no longer present in the experience. We also leverage different-colored text to indicate various meanings. You can come up with your own key, but we have leveraged blue in the past for future design, not yet approved for implementation, and red for items where we need additional information to finalize design.

As the developers go through this documentation for updates and new design implementation, they will be able to refer to what the original design was, which will help them find the location of the component within the project development files. For example, if you have an image that needs replacing in a pop-up panel, seeing the former image as well as the new image will help quickly pinpoint the where in the project that change needs to occur (an example of something we would call a hot swap). The same goes for interactions. If we originally had documented that the delete object tool performs a visual indicator for 4 seconds, but it changed to 2 seconds, we would strike through "4 seconds" and type "2 seconds" (~~4 seconds~~ 2 seconds) so that the developers see the old implementation and the new, annotated correctly. You can create your own annotation standards for design documentation updates—each team works a little bit differently.

Tasking

Once your design documentation has been updated, you can incorporate this into whichever task-tracking tool your team uses. In the production chapter (Chapter 7), I mentioned that my team uses Jira, which supports an abundance of customizable options. You can set up your task cards to support a variety of fields and tags, so use those to your advantage as you set up the workflow for your team. When we have design changes in a project, we typically classify those in a different way than we would a typical task card. We either nest them all within a "story" so that all design changes in a section of the project are under one main card, or we leverage a common tag to indicate that this update is affiliated with a new round of design.

I commonly hear developers and artists ask our production team for as much detail as possible in the task cards when making changes to the original design. The updated requests are typically nuanced and require more specifics than the original tasks, which had some flexibility. Now that we know something wasn't working and need to make a change, communicating that change clearly will directly lead to proper implementation. It can be helpful for the developers or artists to have a working session with the designers or even test session moderators to ensure the intent of the change is carried through to implementation.

After you've worked through the items you've identified for change, updated your design documentation, and created task cards for the production team, prioritization is key. There may be an overwhelming amount of feedback or changes, and identifying where to start and which elements are most integral to the application will provide direction for the remainder of the project. When working on any extended reality experience, it can be difficult to rein in the scope, especially when the technology supports a "sky's the limit" vision. However, by considering the impact on the user experience and original goals of the content, as well as limiting project factors such as time and budget constraints, those priorities become clearer.

User Experience

Creating a good user experience for an extended reality application is a requirement. Immersive technology is designed to be, well, immersive, and so without that, it loses its purpose. The user experience can be affected by the interactions themselves or by the user flow, which brings sense and direction to the content. If there were points in the experience where the user had to stop due to confusion, resolving those issues and redesigning those components should be prioritized over visual polish or minor interaction changes. If the tester didn't know how to perform an action due to instructional content, that's integral to the application's success as well. Focus priorities on elements that would be required for the user to get through the playable loop smoothly and without frustration.

Comfort is also a required part of providing a good user experience. If you have issues with frame rate or general comfort, those need to be addressed as the highest-level priority. Issues with comfort can render a program unusable, which ruins an entire deployment. If any of your tasks in this iterative process focus on user comfort or optimization, bump those to the top of the list. Ensure that when you redesign any of these elements that you are doing so directly as it relates to the hardware the end user will leverage. While there are some universal standards, confirm that the production team is making comfort updates specific to that hardware. If any outside haptic devices are integrated in the experience, consider those as well and ensure all parts are working together as expected.

Content Goals

Think back to the original goals of your content and why you were creating this experience to begin with. If the extended reality experience is training focused, prioritize which iterative elements will make the biggest impact to those organizational goals, which likely focus on knowledge retention and job process proficiency. For your trainees to understand the subject matter, content updates may be the most significant to the overarching application. You would have likely included subject matter experts in your test session if you are producing a training experience, so focus your priorities on elements that didn't align with the educational material or their expectations.

If you're producing a product visualization tool, the goals of your project are likely focused on visual quality of content and accuracy of the placement of that content, which ultimately will drive sales of your products. Yes, interactions are important as well, but the spatial placement and scale of digital objects should be the priority when determining what updates make it into the final application. If your project is focused on an experiential marketing or entertainment experience, focus largely on the level of presence your audience feels. Is the content engaging and immersive? If you receive feedback that it's not, correcting that and making content changes is your biggest priority.

Time

Time is something that affects all projects, not just extended reality. When prioritizing what changes you'll make in your iterative process, remember the hard deadlines you have set before kicking off development. Verify that those remain the same as anticipated, and plan your priorities to fit within the original dates. Focus on integrating the updated design tasks that align with your user experience and content goals first. Then list out the additional "nice to have" changes that your team would like to make. In most task-tracking software, you can assign a priority level to each card. Use this so that it's clear to your developers which items are a requirement within the original timeline and which are good to incorporate after those highest-priority items are complete.

It's possible that your stakeholders will want to extend the timeline to add in new features or components that were discovered during the user testing and iteration phases. This happens often in our projects because once our clients see it, they get excited and want to add more. We try to encourage them to stick with the initial scope as designed and within the original timeline. We provide an option to add a phase II or software update shortly after initial deployment to incorporate some of these requests. If they determine they want to extend the timeline to accommodate additional features or more complex, lower-priority changes, we ensure that the impact to the schedule is properly communicated so that there are no surprises when that original delivery date arrives.

Budget

Budget is similar to time in that it is likely something that was established at the beginning of the project. It's technically subject to change, but it's unlikely to change without a good reason. When establishing your priorities, consider the budget of your project, how much you've already used, and what these iterative changes will mean for the hours and resources required to complete these changes. Ensuring that you complete all requirements as outlined in your original scope is most important when producing extended reality content for a client. We try to account for a portion of the budget to go toward iteration and unknowns that come up throughout the process, but if you don't keep an eye on this, it can get eaten up quickly.

Going back to your stakeholders and asking for additional budget can be challenging, especially when they have expectations of how much something should cost. If you are going to do this to incorporate design changes, ensure you have good reasoning and documentation for this request. If feedback comes out of your test sessions with real end users and implementing their requests is going to cause production to go over budget, that's a discussion that needs to happen as soon as it's known. Stakeholders (whether they are from an external client or internal leadership) should be able to help the production team prioritize in this scenario. They will be able to weigh the benefits of expanding the budget and accommodating the user's requests or deciding against it. Regardless of whether you're sticking with the original budget or getting additional funds, align your priorities and task management to fit within your budgetary constraints.

One of my closest advisors and friends would always tell me, "Perfect is the enemy of good." While striving for high-quality standards is integral based on the technical requirements needed for extended reality production, perfection can be fatal to the product. The iteration process can drive someone crazy if they're always striving for perfection. This is the challenge when the medium in which you're producing the experience will always have at least some subjective elements to it. Nothing is going to be perfect for everyone. This is why knowing when to stop is important. It may be a team effort to help each other identify when something is good. When you are so close to a product, it's easy to keep working and working on something until you think it's perfect. Polished development can waste precious time and distract the team from the true priorities when moving forward.

12

Development

WHILE THE PROTOTYPE chapter included many preliminary steps of development, getting deep into the details is best saved for after initial user feedback. While I'm not a developer, I've observed my team over the years, as well as helped prepare technical requirements and facilitate that between our clients' vision and developers' capabilities. With this experience, I know what's realistic for the development team, as well as when it's time to cut features. If you find yourself in a production role, try to observe and learn from the developers as you go so that you can advocate for them when they aren't in the room. Be realistic about functionality expectations and the time it takes to produce these complex experiences when having discussions on their behalf.

I want to use this chapter to also demystify developer terms and practices as a non-developer myself. There are many helpful tools and methods that they use to produce augmented and virtual reality content, so it's important to have a baseline understanding of these. Some of these tools may be required for every project, whereas others are specific to the technology, hardware, or functionality you're trying to achieve.

SDKs

Software development kits, or SDKs, are leveraged with almost every hardware platform you would use to deploy an extended reality experience. Each headset manufacturer is likely to have their own SDK. For example, Meta provides their Oculus Integration SDK, which can be integrated with Unity to support app development for their hardware. It includes a variety of resources and helpful tools (including libraries and packages) specific to developing for their headsets. There is an SDK that most XR

147

developers leverage called OpenXR, which is a royalty-free open standard (meaning anyone can use it) that works with most XR devices. This means that we don't have to use each headset's specific SDK anymore. We can leverage OpenXR to make it easier for cross-device development.

If your project is going to leverage an external haptic device, such as a hand tracking peripheral or haptic gloves, there will be an additional SDK that your development team needs to integrate into their working files. Like the headset SDKs, the manufacturer of the haptic device should provide you with the SDK included in your purchase. In many cases, hardware manufacturers will release SDKs before the release of their hardware so that developers can get a jump start on integrating and supporting their devices. This is also a great way for those manufacturers to get early feedback before launch.

Libraries

Libraries are collections of code, or functions and classes if you want to get more technical, that help developers produce their applications efficiently. Libraries can be programmed by the development team themselves, but it's more common to start from public libraries, either licensed or, more commonly, open-source. When I mention that libraries are efficient, this means that libraries can make system calls instead of your programmers manually coding those calls multiple times in every instance they are needed.

Libraries are often hosted in a public repository such as GitHub and can be accessed from these repositories and referenced in your projects without manually adding them to each project. Because of this, if an update or improvement is made to the source library, those updates can be carried through to all the applications that reference it without making a manual change.

Assets and Packages

Assets and packages allow for specific customization, typically a feature or add-on functionality on a per application basis. Unlike libraries, assets and packages will be integrated into each individual project. There is a vast collection of XR-specific assets and packages available in the Unity Asset Store and similarly, available in Unreal Engine's Marketplace. You can also produce your own assets and package them for future project use or to license to other extended reality developers. These will often serve a single purpose and save the developer time by not requiring them to code the functionality of the feature themselves.

One asset we recently leveraged had to do with the movement of ropes. Creating rigid objects or even objects that have moving components is well within our wheelhouse. However, we knew if we had to program the flexibility of a rope and how it reacts to any object that it could intersect with, this would take more time than we had available in our development cycle. We were still interested in adding the feature, so we opted to license an asset from the Unity Asset Store. After some integration and tweaking, we were able to incorporate flexible reactive ropes into a workshop scene in one of our virtual reality projects. If there is a very specific feature or issue you're working on, give those asset stores a look to save you time.

In that last example, we were lucky to find an asset and integrate it within our project timeline. Unfortunately, that is not always the case. Knowing when to cut features and how that will make an impact on the overall project is a skill to have not only for your production team but also your developers. Knowing when to call it is important in the iteration phase, and when you're reaching a deadline, sometimes concessions must be made. Thinking of your stakeholders and end user audience can be helpful when making these judgment calls, so reminding the production team of that, especially near the end of the production cycle, can help wrap things up and put a bow on it.

Visual

Putting optimization aside, there are some other visual concessions that can be made when nearing a deadline. However, I would say that sometimes the visual elements are the most noticeable, so gauge the time you have remaining with the tasks at hand and prioritize the obvious visual gaps first. When putting finishing touches on the visuals, you want to identify anything that may still be distracting to the user. There are two oddities that occur in XR that you should look out for—Z-fighting and aliasing.

Z-fighting is when two objects are sharing the same plane and the program can't decide which one to display in front of the other. This causes a flickering anomaly that can not only be distracting but nauseating to the user. This can be resolved by moving the objects in a way that this interference doesn't occur. It's an easy fix and can make a huge difference visually.

Aliasing is the other visual distraction, which looks like the edges of your objects are shimmering and not in a good way. Most development engines would have a feature called "anti-aliasing" turned on. The challenge with turning on anti-aliasing is that it has the potential to lower your frame rate, which can cause discomfort to users. If you encounter aliasing with your models, make sure you are balancing the settings of your anti-aliasing tools to reduce the shimmering but not lower the frame rate too much.

Interactive

When finalizing interactivity in your experience, focus on the elements that have the highest impact to your user experience. This means anything that influences the completion of your core playable loop needs to be polished before you finalize anything else. Keep an eye out for edge cases as you develop, as well. We were recently working on a lockout/tagout process virtual reality training experience and noticed something that only happened for one of our team members. If they pulled the lever to turn off the machine, the lever would float back up ever so slightly, and when they went to put the lock object on the lever object, the lock fell to the ground. We found that this individual was performing the action while sitting, and the angle at which they pulled the lever down didn't align with how that action was performed standing up. If you can't place the lock in this experience, you can't complete the task, which is a big problem. We were able to identify this and update it as we polished the interactivity. Try to break everything as you finalize your implementations, especially interactive features. You never know what oddities you might find.

When considering interactive polish in development, don't forget about the user interface elements that allow your users to navigate. This can include start menus, settings panels, and in-experience options or pop ups. These likely have an impact on the successful completion of the playable loop, so ensure they are connected smoothly and function as designed. These menus may also include options to take screenshots, change locomotion methods, adjust audio, enable accessibility settings, save an experience, or close the application. Ensure that all of those features are functional and that they save across sessions, if applicable.

Add-Ons

Once you have developed and polished your core visuals and interactivity, you can focus on add-on features. It's possible you've already done some design work and prototyping of these features, but these are the components that could be cut in a crunch. Don't work on refining these features until you are certain all requirements can be completed within the timeline. Add-ons can include quality-of-life functionality. For example, I can get along just fine in my virtual gardening game by bending or reaching over and grabbing decorations and plants by touching them directly. However, it would be much nicer and faster if I had a "distance grab" feature that allowed me to grab items at a distance so that I don't have to bend over every time I want to drag my garden gnome five inches to the left. We ended up implementing this distance grab feature as a wonderful quality-of-life improvement, but it's not essential to the core experience or the ability to complete the playable loop.

You also want to ensure that implementing these "nice to have" features doesn't introduce new bugs or integration issues with other required components of the experience. As your development team works on these, it's common (and recommended) that they keep this work isolated in a separate branch of their version control software so that the stable code isn't affected. Once it's verified that the feature is solid, that code can be merged into the rest of the project. We've used these add-ons as incentive for the development team to finalize and polish required features. In the iteration chapter (Chapter 11), I mentioned it's great to know when to stop and when something is good, but it's hard for developers to see that for themselves. By documenting but holding off on the final implementation of add-ons until all the requirements are complete, it incentivizes the production team to prioritize well and be efficient with their work.

Metrics

One factor in finding and defining our use cases is determining your anticipated return on investment. There is no way to successfully measure this after launch without metrics and analytics to determine key performance indicators that are programmed into the experience itself. Some deployment platforms, which we will dive into in a couple of chapters, offer a dashboard with some limited information. However, if you want to understand all the actions your users are taking, programming analytics into your experience is the best way to gather that data. This process typically occurs closer to the end of the development cycle, as all interactions need to be completed before finalizing analytics implementation.

Events

Similar to analytics tracking that websites or mobile apps use, each action that a user can take is referred to as an "event." Events can be recorded as any action or sequence of actions someone takes within an experience. For example, picking up objects, releasing items, connecting an object to another object, entering a room, and flipping a switch are all examples of actions that can be tracked as events. If you're producing a training experience, it may be important to log whenever the trainee completes an action as a part of the process they're learning. This is helpful to confirm that a trainee can successfully complete their job tasks, proving that they are prepared for their job in the physical world.

Events tracking can be helpful for product visualization tools as well, especially when sales goals are involved. By identifying which products your users are looking at and interacting with, you can gather valuable information about what your customers are most interested in and leverage those insights to promote the most popular products. Evaluating analytics and events data can also help you identify when something isn't working in a deployed experience. If you notice users always get to a certain point in your experience and then the events tracking drops off, there may be an issue or hurdle your users are encountering. When you notice a pattern like this, go back and test the experience again to identify if it's an issue with the events tracking or the application itself.

Heatmaps

Heatmaps are used to understand what your users are looking at in a virtual environment, in full 360 degrees. Where events are logging a single action, heatmaps are an overlay of the points of interest in the user's line of sight across an entire environment over time. Think of a weather or radar map and how the areas with the highest amount of precipitation appear red, which then gradually fans out to yellow, green, and then no color at all where there is zero precipitation. The same visuals are leveraged for an extended reality heatmap.

The headset is tracking where the user is looking at all times, so by leveraging this positional data and tracking the user's gaze, you can form a heatmap. Areas where the user looked frequently would show up in a color affiliated with the highest concentration of views. If the user only looked at an object a few times, this would show up in a different color, and for locations where the user never looked, there is no color overlay. When setting up heatmaps, you can create your own key regarding what colors you want to use. There are also services or assets that can be leveraged to set up heatmaps within your project.

XR-Specific Analytics

Another type of analytics that you can track, which are special to extended reality, are positional and movement-based tracking. Heatmaps leverage gaze-based data, which would be considered XR-specific analytics. In addition to gaze-based tracking, you can also track other head movement, arm movement, and even eye movement when using eye tracking supported hardware. By leveraging these methods of tracking, you can gather experiential data and understand how someone is moving throughout the system. It's much harder to cheat the system when this type of tracking is enabled. When I used to play Wii Sports, I'd get lazy. Instead of bowling with full motions, I became proficient at flicking my wrist and still getting a strike. This wouldn't work when movement-based tracking is implemented in extended reality because you're not only tracking the roll, pitch, and yaw of the controller but also the positional movement through space on the z-axis.

One of our projects requires the user to look both ways before crossing a street. We're able to use the trainee's head position to verify if they looked left and right before attempting to cross the street. This is programmed as a tracked event that we can leverage to verify the user performed this part of the process. Based on their success or failure of this checkpoint, we change their experience moving forward. Similarly, we can use the movement of controllers to provide outcomes. In a physical therapy experience, if the user did not hold their arm position for the appropriate amount of time, they did not pass that section of the experience. If they did correctly hold their hand within the target box and keep their head level, fireworks would shoot off to celebrate their success.

Once all your features and functionality have been polished, it's time to build again! When creating builds, it's good practice to ensure your team has a consistent set of standards that are abided by for every build. This should include consistent versioning, release notes, and central location for the files to be stored. Whether building on someone's development computer or a central server using an automated build process, building will take time, which can vary based on the size of the project, so make sure to set your expectations and be patient. We created a PCVR experience for architecture visualization, which meant it was a much larger, higher-fidelity project. Because of the complexity and large environments included in this experience, it would sometimes take over an hour to build. Mobile extended reality projects should not take that long.

Versioning Standards

There are a variety of ways that developers set up their version numbers, so choose what works best for you and stick with it throughout the entire project. Our team leverages a fairly common format that represents each number as Major, Minor, and Patch. For example, if a release had a version number of 1.4.2, 1 = major release, 4 = minor release, and 2 = patch release. When we are doing internal testing and don't want to increment this number up, we sometimes leverage a parenthetical at the end of the version number to represent a hot fix. For example, if we had

1.4.2 (3), (3) = hot fix release. We never release a final build to a client with a hot fix parenthetical. This is more for our quick iterative internal builds.

As you work through tasks in your task-tracking software, your team should always tag cards with the appropriate version number so that the production and test teams will know which version that feature or fix was implemented into. This also makes it easier to pull reports or set filters to see all tasks that were associated with a particular version of the application.

Release Notes

Creating release notes for each build is helpful for the production team to see the status of the project, and it is essential for the quality assurance test team to know what they're testing. Coming up with a consistent format to document your release notes will help with expediting the process and providing a clear overview for the rest of your team. Our release notes formatting includes four sections—important highlights, concerns or scope limitations, release version completed tasks, and known issues. Each of these sections should be filled out prior to handing off the build, and ideally known issues would be empty (or nearly empty) upon final delivery.

Important highlights should focus on the key items you'd like to call out for the release. We format this as a bulleted list and call out high-level features such as "throwing mechanics implemented" or critical bug fixes such as "voice-over trigger resolved." Following that section, concerns or scope limitations is where we would call out any challenges we are having in the development process. This will help us communicate with our project stakeholders when something is taking longer than anticipated or if implementation isn't working as designed. The next section, release version completed tasks, is a list of all tasks that have been tagged for this specific release. We document our release notes in Confluence so we can directly pull a filtered list from Jira, which means no manual entry. I highly recommend you do the same if possible as it speeds up the process of creating release notes. Finally, the known issues section is also implemented as a list, which we have an automatic filter produce. This section notifies your testers of what issues you've already identified and are in the process of resolving. This will ensure there isn't duplicate entry when it comes to bug-tracking cards.

Once you have a completely playable experience, you can start creating builds for your quality assurance team. This should wait until the experience is somewhat polished but before it is finalized. This team should be able to start working through tasks your developers consider complete, verifying their completed status before final polish. By keeping versioning consistent and task cards tagged with the proper release version, as well as providing thorough release notes, moving into the quality assurance testing phase will be a breeze.

13

Quality

QUALITY ASSURANCE, OR QA, testing is a tedious but necessary part of producing any technical product. Where you tested for usability and experiential feedback earlier in the process, you will now test for technical or functional issues with the extended reality experience. Some organizations have a dedicated QA resource or team, and other smaller teams must lean on each other to perform this type of testing in addition to their primary role. Regardless of who is testing, ensure that it's someone who was not directly responsible for developing the sections of the project that they are testing. Each component should require a second, if not third, set of eyes before passing this review process. QA testing will likely go through multiple rounds of testing, especially if you're leveraging new and untested features and functionality.

Bugs

Bugs are the most common thing you'll find during your testing phase. They present themselves as a fault or error in the experience that testers encounter while attempting to complete the playable loop or use accompanying features. Any issue you encounter when testing the experience is likely classified as a bug, and as you document these, make sure to gather as many details as you can to help the developers pinpoint the issue and brainstorm a solution. Some bugs can be funny. One time we had a canister in an experience that continued to infinitely replicate itself as soon as it was opened. After about 10 seconds, the ground was covered in these canisters and the frame rate dropped to an extremely low value due to the massive number of objects in the environment that had suddenly appeared.

When producing content in extended reality, bugs are all around you (that sounds kind of creepy, I know). Look at your entire environment and ensure that you're not missing a corner or section of the space. It's much easier to test for bugs on a constrained 2D screen such as a phone or tablet. But in extended reality, we must observe the entire virtual environment, or overlay of digital content onto the physical world, with full spatial testing. Move and walk around to ensure you're taking advantage of the full space available, as movement can trigger bugs to appear as well.

Locks and Crashes

When the program locks or crashes, that is serious. The player experience is immediately ruined if that were to happen to your audience. We typically refer to a lock that happens in the middle of the experience as either a softlock or hardlock. Softlocks occur when a user gets stuck within your experience. This can be due to the design of the experience and a gap in your user flow where the individual can't progress because of an action they performed. Or it could mean a technical glitch in the experience that makes it where they can't progress. A more serious lock is called a hardlock, which likely requires a reboot of the program and is almost always a technical glitch. If your program has save files, those may be lost or unusable in a hardlock situation. Both types of locks can be frustrating to a user and when identified, should be resolved as a high priority.

A crash is another form of serious bug. If your users are going through your experience and the application crashes, they immediately lose their progression and must start over regardless. In virtual reality, this can be especially disorienting because the system commonly freezes the graphics, which can throw you off physical balance. If your experience crashes, note at what point this occurs, try to replicate it, and identify any actions that were taken to cause the crash.

Technical Function

Perform quality checks for every technical function and interactive element present in your experience. If possible, test the interactivity for each object that supports the functionality you've designed. While performing these checks, ensure that the interactions you've programmed are working and that the controls mapping has been set up properly. For example, if you set up the grip button to function as item pick up interactivity, the grip button should work correctly for all objects. After you've verified the proper controls are good, note the outcome of the interactivity and if there is a discrepancy with the design; that could be identified as a bug. If you encounter an interactive bug, try to get video footage of the action being performed, as well as the incorrect outcome.

Technical functionality can also drive back to cause-and-effect scenarios your team has programmed. In our gardening experience, depending on what plants you've grown, those plants would attract various critters. We set up a matrix to determine the combination of flowers, fruits, and vegetables that were required to attract an animal, as well as several detractors, whether they were other plants or decorations. We had to go through the experience and perform checks to ensure that all our setup worked during this quality assurance process. Because the base system was implemented to work the same for all plant/critter combinations, we performed spot checks of various combinations to ensure that technical functionality was executed correctly.

User Flow

When testing for the user flow, you'll want to reference your original design documentation. Confirm that all paths that could be taken are set up as intended. If there is a deviation in the path that results in an intentional experience failure state or would loop back around to an earlier point in the experience, ensure those paths are functional as well. As you are going through the user flow, be cognizant of softlocks as these can appear as a result of the user flow not being set up properly in development.

We commonly record video of a "happy path" run-through of the experience, which only includes the good outcomes and the smoothest user journey. Consider the happy path as a way the experience can be completed when everything goes right. We start with this as a baseline and then once we verify everything is correct with the happy path, we try to deviate and test all other potential outcomes, whether positive or negative, that have been programmed. After that testing is complete, we will try to branch outside of the user flow and ensure any actions that we perform to "break" the system don't cause issues.

Visual Issues

Visual issues with the experience can be both distracting and unsettling. We already touched on aliasing and z-fighting, which you should always be on the lookout for. There can also be visual issues with objects, characters, and animations. Refer to your original storyboard and design documentation to verify that all visual elements have been implemented as designed. This is especially important to verify for character animations as the documentation and how it was interpreted by the animator can vary. In the storyboard of a training experience we produced, it was written that one of the characters gets mildly injured. The animation was set up in a way that didn't make a big enough impact to even notice that she had gotten injured at all. Remember, actions and reactions sometimes must be larger than life to be noticeable in virtual reality. We identified this issue in a quality check and the task was sent back to the animator to exaggerate the character's movement.

While the animations and art can be subjective and open to interpretation in production, there are some visual issues that are just bugs, plain and simple. We produced a frog in one experience, and everything looked fine until he started ribbiting. Somehow his rig, or the joints that make up the structure that can be animated, stretched to enormous size every time he croaked or hopped. His legs would get as long as a human's legs, stretching the 3D object of the frog into a terrifyingly disfigured creature. Those will be more obvious to note, but the bottom line is that if it looks odd or distracting, it likely doesn't align with the original intent or design and therefore should be noted as a bug.

Scoring and Analytics

The programming behind scoring and analytics must be tested upon the completion of each run-through. If you have scoring built into the experience, you will ideally be presented with that score at the end in a panel to show your overall outcome as well as a breakdown of all elements that went into that score. As your testers are going through the experience, encourage them to remember or write down which path they took to get to the end so that they can verify the scoring is calculated as expected. You can also set up test cases, where they go through a list of predetermined paths to verify the scoring is correct.

Depending on how you've implemented analytics, you'll likely need to access another platform to test whether this information is being passed through properly. Whether it's being logged into a database that you control, sent to an external platform, or integrated with a learning management system, complete your test run-through, and go directly to the appropriate capture system to verify events or scores are tracking. If you notice that the analytics are not being delivered to the system, note whether the issue is that no information is passing through, or whether the information that is being collected is presenting incorrectly. Like the process for scoring, note the path that you take in your test scenarios so that you can match up the outcome with the expectations.

Project Requirements

During the quality assurance testing phase, we must verify that all project requirements are correctly and completely implemented. If you had enough time to add stretch goals or "nice to have" elements within your experience, you should confirm these have been incorporated as well. All elements that your stakeholders expect to be included in the final product should be verified and pass the quality assurance process. This includes interactivity, environments, specialty objects, characters, and content.

When verifying the content has been implemented properly, we will bring back in the subject matter experts we spoke with early in the design process to help with this verification. While getting these individuals into the headset is ideal, that may not always be realistic due to busy schedules or hardware access. If that's not possible, we typically share the "happy path" video we captured during the user flow quality testing process as it helps show them the near-final product so they can do a visual check and confirmation that the content is accurate. This is especially important when producing educational or training content, as we must verify the extended reality adaptation of the material is aligned with physical world expectations.

Run-Throughs and Checks

While going through the full experience in great detail during your initial quality assurance testing round is integral, you may not need to do the full pass every time. Use the release notes provided by your development team as you go through the experience in each iterative build. Starting with the second-round quality build, we typically begin with a spot check of the items that have been marked as integrated or fixed and then move on to a full run-through of the experience. If we anticipate multiple iterative builds in a week, we may save the full, rigorous run-through for the end of the week and use the intermediary builds for spot checks only.

Spot Check

We typically perform spot checks on small items that have no impact to the underlying code or systems used to produce the experience. This can refer to a visual change, an asset swap, a text update, or minor interactivity tweaks. For example, the delete action in our creator tool had a request from testers to change the visual outcome of that action from four seconds to two seconds. This is a great candidate for a spot check. The interactivity itself hasn't changed, but it has been tweaked to result in some change that still needs to be verified in a build. The tester can go into a build and spot-check that the delete action indeed results in a visual outcome that only lasts two seconds.

We commonly must swap out text or colors near the end of a project, especially when our clients see the build for the first time in its near entirety. They know their content and subject matter better than my development team, so they can identify small tweaks that we would never know to look for. This happens when there is a good bit of instructional text in an experience or specifications for a product visualization tool. In an augmented reality visualization tool, the 3D models and placement would have been finalized long ago, but the product descriptions or specifications can easily be modified and tweaked all the way down to deployment. This is an easy change that requires a text swap and can be verified in a quick spot check.

Full Session

A full session is something that will occur multiple times throughout the finalization stages of your extended reality production cycle. When the product is considered complete, run through that full session test and identify any remaining bugs, visual issues, scoring inconsistencies, and misalignments with design documentation. Hopefully you don't have too many items to note, but now is the time to get nitpicky. The full session should allow your test team to go through all task-tracking cards that have been marked with the quality assurance, or QA, status. This status indicates that as far as the developers or artists are aware, the implementation of that task is complete and correct.

When going through the full session, leverage the user flow to ensure all paths are taken. Then try to break things. While verifying all items are functioning correctly, you must also go into the minds of your users—the individuals who don't know the limitations or scope of what you and your team have programmed. Try to mess things up and see what happens. It's possible that nothing will happen, and the rails of the experience will keep you on your path. It's also possible that you'll encounter softlocks, hardlocks, or even crashes during this exercise.

Multiplayer Test

Multiplayer testing will only apply to experiences that have multiple devices networked together. This can be a set of the same extended reality devices, or various devices in an experience that supports cross-device compatibility. When testing multiplayer functionality, take note of your network connection speeds to ensure that any issues you encounter are from the program itself and not your Internet. Depending on what multiplayer features you've incorporated into your experience, you'll want to verify that users can see each other, hear each other, speak to each other, and interact with each other. Multiplayer testing takes significantly more time and resources than single-user experience testing as you must gather several people together at a common time to perform these tests, so plan accordingly.

There may be circumstances where you have designed an asynchronous experience, or a program where the extended reality user has a differently designed experience than the other networked individual. Be attentive to any bugs or issues and how they react to each other. For example, we produced a manufacturing experience where trainees are in virtual reality learning about a machine and the instructor controls their experience from a tablet. The trainees are in a multiplayer setting in virtual reality headsets, can see each other, and can independently interact with the machine. The trainer has a tablet and can make selections to moderate the experience in a guided mode for the entire class of trainees in virtual reality. This adds complexity to testing as the virtual reality users are experiencing something completely different than the trainer, and depending on what the trainer does, the trainee's experience is affected. If you decide to implement multiplayer in any experience, plan for an extended testing timeline.

Integrated Hardware

Integrated hardware is another factor of testing that will not apply to every experience. If your project leverages any peripherals outside of the base headset and controllers, those integrations and the functionality of the devices must be tested. This can include hand tracking hardware, haptic gloves, omnidirectional treadmills, customized controllers, simulator chairs, or any other device your production team has chosen to use. Confirm that the SDK has been integrated into the project and that your developers have these external devices functioning before starting this type of testing.

As you test your external peripherals, ensure that all connections are functional and the devices are recognized by the base hardware. Depending on the device, this could be a wired connection or wireless via Bluetooth or some other method. Confirm that all expected behavior associated with the device is working. If you're leveraging a hand tracking device or haptic gloves, verify that all interactions are functional and no actions require the controllers. If you're leveraging a haptic vest or simulator chair, confirm that it's reactive to all interactive elements that are programmed to have a direct impact on its functionality. Pass through a full session, including the happy path and the deviations from the user flow to ensure nothing is missing or broken.

When issues are identified, document them with as much detail as possible so that your development team has clear direction on the issue and can figure out how to resolve it. You will incorporate this detail into the task-tracking cards so that they have direct reference of how to replicate the issue. If the issue cannot be replicated, still document it, but lower the priority or create another categorization. It is extremely difficult to fix a bug that can't be replicated, because you have no real way of testing to know if it's resolved. It's like taking a shot in the dark.

Steps to Replicate

Documenting the steps to replicate a bug is going to be the most helpful deliverable when creating task-tracking cards for your developers. Some task-tracking systems have special formatting for bug documentation that includes a section specifically for steps to replicate. Take advantage of this if it's part of the system you use. When writing down steps to replicate, start as far back as you can in the section where the bug occurred. Document these steps in a bulleted or numerical list, and end the list with the outcome, lock, crash, or bug that you experienced.

For example, when going through a full session test of an augmented reality application, one of the 3D model interactive buttons was not functional and then proceeded to crash the app. This would be documented as such:

1. Open the product page.
2. Click the "View in AR" option.
3. Place the 3D model in your space.
4. Tap the transparency icon.
5. The app immediately crashes.

While this list may seem like an overly explained path to getting to the crash, it contains step-by-step instructions for the developer to see the bug for themselves. They may not have the ability to communicate with the tester or ask follow-up questions, so it's important to put this information in the task-tracking card. Testers should always verify whether they can replicate a bug consistently or inconsistently before passing the task off to the development team. If it only occurs 50 percent of the time, note that as well.

Reference Footage

Obtaining reference footage of a bug is a gold mine for the developers as they try to resolve the issue. Ideally, the steps to replicate documentation should be enough to trigger the bug themselves. However, if you can provide reference footage, they can more quickly get to pinpointing the issue and coming up with a resolution. We don't record every session, but if we encounter a bug, we record footage of ourselves replicating the bug and attach that clip in the task-tracking card.

Some testers want to record every session each time, which is completely fine. This leads to massive amounts of video footage, so if you choose to do this, come up with a consistent way to label the footage and a central location to save it for all team members to access. I find that the happy path footage and then specific bug, lock, or crash footage to pair with the steps to replicate documentation are sufficient.

After your extended reality application passes your internal quality assurance testing, it's ready to be delivered. We initially provide a limited release only to stakeholders, whether internal or external, in a phase called user acceptance testing. It's during that phase that all the requirements are checked off from their perspective and we can close the project. They may find issues we didn't find, at which point we go back to the development team and start the process all over again. Once everything passes user acceptance testing, it's time for deployment. The fruits of our labor will finally reach their target audience.

14

Deployment

It's the moment you've been waiting for—deployment. Just because the product is ready doesn't mean the extended reality production process is over. There are still different steps that must be taken to have a successful and effective deployment. You want to make sure your project files are preserved so that if you need to go back and edit or update them in the future, they're available. You also want to make sure that you're deploying in an organized way. Documentation has come up as a common theme throughout this entire production process, and deployment documentation is no exception.

Packaging

When your product is ready for deployment, you'll want to package your working files and archive them as a backup. Depending on who you are producing the extended reality experience for, your stakeholders may also require a backup copy of the packaged files as a part of your agreement, so determine that prior to starting any work. Before you give the package to anyone outside of your production team, confirm who should have access to it as it may include some of your proprietary code or intellectual property. Any software you've been using to create your extended reality application should have the option to package up your working files.

This packaged format will include all files and references associated with the functionality and visuals of your project. It should include many folders and subfolders, and the file size will be much larger than the built product itself. This package will include all 3D models, as well as their accompanying textures or texture atlases, and any plugins, SDKs, or libraries that your project references. The idea behind the packaged files is that a new developer should be able to open the package on their computer and have everything they need to pick the project back up for an update.

Version Control

In the development chapter (Chapter 12), we reviewed versioning standards and the meaning behind the format Major.Minor. Patch. When you're nearing your first deployment, consider what you want that structure to be and what number you want to launch with. Typically, an initial launch is 1.0.0, although if you're doing a public early access or beta launch, you may want to keep that first number, or "Major" release number as 0 until the official full release. If you do want to launch with a 1.0.0 release, keep that in mind during your development process. Start with 0.1.0 and roll up from there so as not to accidentally interfere with your release versioning.

You'll want to make sure that you have consistent standards for your version numbers and that they're aligned with your version control software moving into your release. It's nearly impossible to roll backward on a version number, and most deployment platforms will not support it, not to mention it would be confusing to your users. We try to also incorporate the version number somewhere into the visuals of the build itself. A common place to add the version number is in the start menu or settings menu. If a user has an issue after release, this helps them confirm which version number they were using so that they can report the bug properly.

Version Code

The version code is specific to Android device deployments. This is relevant here because most stand-alone virtual reality headsets, such as the Quest and PICO series devices, are Android devices at their core. When publishing an Android PacKage (or APK; odd that they decided this would be the acronym), most platforms will reference and require the build code of the experience to be incremented up. This number must be an integer (whole number), and it needs to be greater than the last integer that was deployed with this program.

If you upload a build to a deployment platform without this change in the project, the platform will reject your build and you will not be able to publish that version. Your developers will need to update the version code, build again, and attempt an upload once more. Remember to update your version number and version code each time you update a build to a deployment platform. Add this into your production process as a step that is required for every release build to get into the habit of doing so. This version code is referenced by the platform to confirm that the build is newer than the previous build.

Deployment Methods

Now that you have all the technical components out of the way, you need to determine where you're actually deploying your product. Hopefully this would have been determined in your early production stages so that your team could incorporate any standards or nuances for each deployment method into the development effort. However, now is the time to make the final decision and start preparing any additional materials, descriptions, or requirements for each of the three deployment methods—limited, public, and private.

Limited Deployment

Choosing a limited deployment is good for either short-term beta launches or if your experience is going to be deployed for a specific event or limited release. If you're using a mobile virtual reality system, the quickest way to get content onto devices for this type of deployment is called sideloading, or directly transferring the build onto the device while plugged into a computer. This is only good if you're doing this for a small handful of devices as a temporary solution. Each time you want to make an update, you must plug the device back into the computer, remove the old files, and transfer the new build. If you do not delete the previous build, the new build will not work. While this can be done with your file explorer, it's much easier to use a free program called SideQuest to facilitate these file transfers.

If you're producing an augmented reality experience for iOS devices, TestFlight is a great tool to leverage to get the early or limited version of your experience out to users. Be cautious if you use this method as each build can only be live for up to 90 days. You must also manage each of your users by individually inviting them via email or allowing them to join via an invitation link. While TestFlight should not be leveraged as a long-term deployment method, it is great to quickly and easily get builds to your beta audience before public launch. You can move apps from TestFlight and submit them for public publishing as well, creating a smoother deployment experience.

With a computer-based virtual reality deployment, or PCVR, you can load the build directly onto the computer as an executable file for a limited release. Be aware that there are likely other programs required to run simultaneously with your experience, such as SteamVR, so install those as well. Similar to sideloading, the launch file needs to be individually downloaded or transferred to each computer that is running the experience, so this method is not one that should be used for widespread launch. However, if you're doing a deployment on PCVR for an event with several virtual reality setups, or a training program that has dedicated space in five facilities, transferring the executable and launching it directly from the computer is your best path to deployment.

Public Deployment

When launching your extended reality experience publicly, you want to ensure you have the requirements for each public platform successfully checked. Each public store has their own set of standards that applications must pass before being allowed to launch on their store. For example, Meta has something called Virtual Reality Checks, or VRCs, which include dozens of checkpoints relating to content, comfort, accessibility, performance, security, privacy, and more. Most stores are transparent about their requirements, so do some research for your preferred store and incorporate those requirements into your development cycle so that you aren't surprised upon publishing. While the Meta Quest store has the most rigorous requirements and review process, you can also publish your program to other virtual reality stores. SideQuest, which is the program I mentioned to support sideloading, also has a public store that is widely used with simpler upload standards. If you're producing a PCVR experience, Steam is the route you will want to take for a public deployment.

For mobile augmented reality applications, the App Store for iOS and Google Play for Android are the most common public deployment methods. Each of these stores will have review times after you submit a build for publishing, so bake that time into your deployment schedule. If you're creating a social augmented reality filter, those can be published via creator tools that each platform makes public to their users. For example, if you want to publish a filter to Snapchat as well as TikTok, note that you will need to go through the formatting and process of developing and deploying to both separately. Regardless of whether you're publishing for virtual or augmented reality, every public platform will require you to have a developer or creator account, so plan accordingly.

Private Deployment

Most enterprise or business applications will never reach a public audience, and that is intended. When producing content that has proprietary material or processes specific to a business, the application often stays internal and therefore needs a secure and managed way to deploy to devices privately. While sideloading can work for the short term or for preliminary reviews with stakeholders, a better way to deploy is via a mobile device management system. These platforms have been around for years supporting devices such as laptops, phones, and tablets. However, in the past few years a small handful of companies have emerged to support extended reality mobile device management and from my experience, it has been a game changer. The platform we leverage is called ArborXR, but there are others out there, so do the research for yourself before choosing the best fit for your organization.

The best part about mobile device management is that we don't have to physically plug the devices into our computers to transfer files. Once the preliminary setup is complete, we use an online dashboard to manage devices, groups, and content in one place. We can also confirm which devices have been updated to the latest version of the content, as well as reboot or initiate content launch directly from our web portal. This is crucial when managing a fleet of tens, if not hundreds or thousands of devices for one organization. When the goal is to extend the content to a large private audience, there really is no other scalable way to do so. As a team of extended reality creators that produce projects for multiple organizations, we can manage each of our client organizations separately and with no overlap.

While I mentioned phone and tablet augmented reality in limited and public deployment methods, extended reality–specific mobile device management platforms can support many wearable augmented reality headsets. Most of these wearables aren't for the consumer market yet, so this is a great way to get the content you produce to your business users. The list of supported devices is always growing, and it's nice to have one platform that supports all of them. These mobile device management platforms also typically provide device information, such as battery life, firmware version, device settings, and even location. This information is crucial in maintaining the health of devices across the organization and ensuring they are in the right hands.

Communication Channels

Whichever method of deployment you choose, you'll want to set up avenues for your users to communicate any challenges, unexpected issues, or feedback to your team. If you're producing the experience for a client, that feedback may be something they want to manage themselves and filter back to your team. Setting up these channels of communication will help you stay in contact with your end users and make it easier to deploy updates that in some part are a direct response to their feedback. The relationship with your end users doesn't stop at the initial deployment, so keep that in mind as you move toward your launch date.

Making Updates

Some updates may be planned with new features and functionality that continue to enhance your experience. However, in some instances, you'll receive feedback from your users who have identified (hopefully) minor bugs and issues that are filtered back to the production team. In the case of these issues, you may want to plan for a patch release shortly after deployment, which would result in the increment of the third number in your version number to increase by one (e.g., 1.0.1). Depending on the severity of the issue identified, you may be able to wait until the next planned update. In more serious cases, especially when the reported issue causes a lock or crash, it must be addressed immediately and inserted into the post-deployment production plan with urgency.

On the other hand, planned updates should be structured into a release schedule so that the production team can start working on task-tracking cards and build schedules. When working on the release timeline, consider complementary features, functionality, or visuals that pair well together. Try to group interdependent components into one update. By planning the updates with some form of structure, this will provide a clear vision for your end users. Some developers want to release planned updates and feature road maps publicly to their audience to grow audience excitement and increase user engagement. If you choose to do that, and especially if you release tentative dates publicly, ensure you can stick to those commitments and your release schedule.

Collecting Feedback

Whether you are going to be the one collecting feedback, or it will be managed by your stakeholders, creating a process for post-deployment user feedback is important. As the most basic option, you can create a simple form to collect feedback through any form builder. Another option is to leverage a tool such as Jira Service Manager, which serves as a public link for your audience. Issues that are submitted via this link are automatically input from your external audience directly into your project task backlog. There are professional services that manage customer feedback as well, which may be of interest if you have a large or widespread deployment.

If you want to create a sense of community or have a more personal connection with your users, which many experiential or game launches choose to do, consider a platform such as Discord to facilitate that community. You can create different channels to encourage conversation and feedback in a positive and constructive way. It also gives your production team a direct method to learn from and communicate with your users, which can lead to more meaningful updates moving forward. Regardless of which method of feedback collection you choose, you want to make it clear to your audience so they don't take their issues directly to a review on your product listing. This would be especially damaging if you have a public deployment.

Documentation

Communicating with your audience is important, especially when you're trying to attract people who have limited exposure to extended reality content. While most of the education about how to use your experience should take place within your tutorials, you can produce additional supporting material to help the end user when challenges arise. Creating a package of documentation, reference material, or video content for your audience may be a great way to help them optimize their experience.

Instructional Guides

Think of an instructional guide like a product manual you would receive with a new kitchen appliance. It can include quick-start information, things to know, and any supporting information that may not be a fit to include within the experience itself. Something we also commonly include are basic instructional guides for device-specific operation and troubleshooting. Often these are available via the device manufacturer, but sometimes we will provide simplified or specific guides for the devices our clients use. These guides include information such as turning on and off the headset, rebooting the headset, connecting to Wi-Fi, adjusting volume, and setting up their room-scale playspace.

In addition to instructional text or images, we often create quick videos to walk through controls mapping and how to use the controllers. This can be especially helpful for individuals who don't have familiarity with gaming controllers as there is some crossover. We will also sometimes include video footage of the experience itself or playthrough footage with narration so that users can reference this if they encounter a challenge or issue. These videos are not meant to replace the experience but provide context and support if someone is unfamiliar with the technology and how to use it.

Marketing

Now that you're ready to launch your product, it's time to let the world (or your internal audience) know that it's time to access it. Plan your marketing strategy well before launch, and have a communications plan ready for launch day and the weeks before and after release. If this is a publicly available experience, some deployment platforms support preorders or wish lists, which will notify your users when the product is ready for download. If you are launching on a platform that supports this, try to get those communications out early to get your product a head start at success from day one.

If you are only launching your product internally or to a limited audience, marketing and communications are still important to the overall sentiment toward the experience. When deploying internally within a business organization, we often see hesitancy from employees before using the experience. This is common and typically can be resolved with education around why this product is going to help them with their job. Communicate why this technology is being introduced and what goals are aligned with a successful deployment. Don't just throw the product out to your organization and expect immediate adoption. Unfamiliar technology can be met with resistance so communicating and educating is sometimes just as important as the product deployment itself for the best chance at success.

Having a successful launch will be deeply satisfying! Enjoy it, and soak in the fruits of your labor. Some experiences, such as marketing campaigns or experiential events, may be one and done. Many experiences will require post-deployment support or updates, so set expectations upon deployment for how users can provide feedback should anything arise. Building on a strong release can lead to a fruitful long-term run of your extended reality product.

15

Measuring Success

YOUR PRODUCT IS out in the wild now, or at least with an external yet controlled audience, so what happens now? We can look back at the very beginning of our process where we defined our use case and refresh our memories on the performance metrics we established. This will help us note how the product is performing so that we can make informed decisions on what updates will be most relevant moving forward. While it's important to have a channel to collect user feedback, we can also gather insights from events tracking the development team implemented in the project, as well as information that our distribution platforms provide or scoring data from a learning management system integration.

Events Tracking

The development team should have implemented events tracking so that you can observe user behaviors as well as unexpected behavioral outcomes after deployment. While the data you collect may not be from one specific user, and it shouldn't be due to privacy laws, you can make assumptions and generalizations about patterns you see in the data. Identify if your audience is going through the user flow as you've designed it. You will be able to see the path of their actions as well as how long they are spending in each trackable area. Document if this aligns with the anticipated user flow or if they deviate from it in common ways. If a group of individuals are making the same deviations, consider why they are doing so. Is this because your users want to take that path as it's a better and clearer experience for them, or is it due to lack of instruction or confusion within the experience?

You want to note whether there is a particular drop-off point where you lose users. If you notice the events or user actions significantly reduced midway through the experience, that may indicate a crash or softlock, or this behavior could indicate loss of interest in the content. When you see this pattern occur in your analytics platform, go into the experience yourself to verify which option may be the case. If it's a crash or softlock, that needs to be resolved immediately with a patch update. If it's the latter, try to gain direct user feedback as to why they may be losing interest and what content improvements would enhance their experience. Alternatively, you could see analytics of users completing all available content in its entirety and returning to the application to experience it again. If that pattern emerges, you may want to consider a series of content updates to give your audience more to consume to keep them engaged.

Platform Insights

Most deployment platforms will give you some general insights into your user data or the hardware itself. This varies by platform, but common metrics include time spent in your XR experience, number of installations, returning users and their frequency, general location information, battery usage, crash reporting, and even some demographic information such as age group or gender. Public deployment methods such as stores or marketplaces will give you data on the users you don't have direct access to, whereas private deployment methods such as mobile device management platforms will give you a way to keep track of exact device and user information.

These insights may not be as applicable in indicating the user experience of the extended reality application but can prove helpful in other ways. For example, on the Meta Quest developer dashboard, I can see how many individuals install and play my public virtual reality experience on a monthly, weekly, and daily basis. There is a significant drop-off from monthly users to daily users, indicating people download my experience, play it, but don't come back every day for more. However, if I wanted to get my daily user count to increase, I could introduce special daily tasks, or a randomly rotating item drop to encourage users to log in each day. When I publish this update, I can see this number grow and confirm that I've reached my goal of more daily users. You may not leverage every metric on these platforms, but get creative with how they can help you learn more about your audience and improve the product.

Other Metrics

You may have programmed other metrics such as scoring data or heatmaps, so include those in your review as you make informed decisions regarding the success of your deployment and what to focus on with your updates. Scoring data can be gathered from a third-party analytics platform or your learning management system, depending on how you've integrated it within the project. Whether it's a game, experiential marketing application, or training simulation, scoring data can educate your stakeholders on how much the audience is retaining the information within the XR experience. If trainees are scoring low, identify why this is happening. Is the instructional content in the simulation unclear? Alternatively, if they are scoring off the charts, this is a great indicator that extended reality training is proving to be more immersive than classroom learning.

If you have programmed heatmaps into your suite of metrics, review those and note any points of interest, as well as areas where there may be a lack of interest. Compare this information to your design documentation and align it with your original expectations. Confirm if users are focusing on the elements you've designed to catch their attention. If you notice that people look over at one specific area that was supposed to fade into the background, go back into the XR experience and see what's so fascinating for yourself. It could be a distracting element such as z-fighting or aliasing. Or it could be that the common user behavior in that part of the XR experience has them exploring a part of your environment you weren't expecting. Leverage this additional information to your advantage when gathering user insights.

Once all the various data, analytics, and metrics have been gathered, we incorporate that with other organizational factors to measure success. There are so many ways to measure the impact of extended reality for an organization that I am bound to leave something out. However, there are some outstanding key performance indicators that I like to measure, including organizational efficiency, safety impact, sales goals, cost savings, and brand impact. By documenting pre-deployment benchmarks and aligning your metrics with post-deployment data, you can calculate the ever-important return on investment, or ROI, that every stakeholder wants to see.

Organizational Efficiency

Time is money as they say, which is why organizational efficiency carries so much weight when calculating a return on investment. Time is measured for almost any job, whether it's how long a task takes to complete or how much time can be saved by performing that task more efficiently. One of the goals of extended reality experiences, especially virtual reality training experiences, is to help workers perform their jobs more efficiently. By practicing their processes and building up that muscle memory in a virtual simulation, they can enter their physical job and be better prepared to perform. If workers are trained more optimally, they can achieve a higher level of success with their tasks from the start, leading to more products being manufactured, for example.

Augmented reality also has potential for a significant impact on organizational efficiency. With on-the-job assistance tools such as operational information overlay for a piece of machinery or line-of-sight picking lists for warehouse settings, employees can have information they need directly within a wearable. By having quick access to this reference material, there is no need to fumble or flip through paper manuals or keep a physical checklist on a clipboard. Regardless of which extended reality application you implement, ensure you are collecting pre-deployment metrics on time spent to achieve product throughput or delays and downtime due to troubleshooting so that you can properly calculate the impact your XR experience has on the organization.

Safety Impact

The impact on safety and the reduction of serious injuries or fatalities is at its core the biggest return on investment you could ever calculate. We're talking about the impact on the life and well-being of your audience, whether it is your employees or the general public. This performance indicator is difficult to calculate and must be done with tact and care. The most obvious way to measure this is to look at historical data and information about injuries and fatalities directly related to your subject matter. If it's an internal deployment, that information may be sensitive or restricted to certain employees. If the XR experience was publicly deployed—for example, a safe-driving simulator—you can research public data on the number of injuries and fatalities that occur in automotive accidents or whatever your subject matter is related to. Compare these numbers to the results you see after launching your XR experience.

Another measurable factor here is aligned with the interests of insurance companies or other financial drivers. Safety incidents can be costly to any organization, not just the human factor but financially. No organization wants a safety incident to occur, as they are expensive both in time lost to halt work when it happens, as well as to pay for any lawsuits, insurance rate increases, or other financial payouts related to the incident. Measuring the success of this may be the longest performance indicator to calculate as you need a significant runway of reporting before you can conclude that your XR experience made a safety impact.

Sales Goals

Measuring sales goals and how they have been facilitated by an extended reality application will most likely align to product visualization tools or marketing experiences. Because sales numbers are already measured, this is one of the simplest to calculate. Identify which products your XR experience supports and view your event data to see how many users engaged with that content. Did they only place the virtual representation of the product in their room, or did they also explore other key specifications and content related to the product? Compare that product's sales numbers from before it was supported in an extended reality application and after. Note any changes, hopefully increases, to add to your overall return on investment calculation.

While selling products is the primary goal here, it's also worth noting that augmented reality can reduce the amount of product returns, which can also contribute to stronger sales cycle numbers. Because users can see the product in their room, they are making more informed decisions and have an exact idea of the measurements and look of the item in their physical space. Make sure that as you calculate the impact of extended reality on your sales goals, a reduction in product returns is included in the mix.

Cost Savings

Cost savings via extended reality can be calculated for several reasons. Savings on travel expenses can apply to both a sales and marketing tool, as well as training experiences. With a sales and marketing tool, your customers can now see a library of products in their own space, reducing or eliminating the need for in-person product demonstrations. On the training side, instead of flying a fleet of trainers all around the world, or all your trainees into your headquarters to complete training, this can be done by sending out headsets. Not only does this save on the cost of flights but also all other ancillary expenses such as hotel and food and the massive amount of time it can take to travel to the location as well. Multiply that by the number of employees that no longer must travel, and you have your number.

There is also potential for a massive amount of cost savings on material waste. With digital models comes unlimited product and reduction of product waste. This can be especially significant if your physical product is used in the training or demo process. You no longer need to trash physical products that are used up and wasted in this manner. Wear and tear on machinery to manufacture your products may also be reduced as training on the machine now puts stress on a virtual machine that can be infinitely reset and requires no maintenance. Calculate the material waste within your organization and incorporate it when calculating your total return on investment.

Brand Impact

If your team has produced an experiential extended reality application, such as a brand or story-driven experience, there may be fewer concrete metrics, but you can still include general feedback when calculating your return on investment. Examine the feedback or reviews you've received from users to gauge the overall perception and sentiment of your XR experience. If you publicly launched on a store or marketplace or created a Discord server for ongoing feedback, look at user engagement within those platforms. Remember to communicate back to the users via comments to let them know you're paying attention to their requests.

Finding glowing or engaging reviews of the XR experience can make a positive impact on your brand. Even if the comments are constructive, that means your audience is engaged enough to say something and connect with you. It will also give you a great starting audience to produce your next XR experience. Building fans of your brand and your content can lead to more opportunities and higher visibility throughout your industry.

Once you've identified your metrics and their impact on the business, you can create new ongoing goals for your long-term deployment road map. Making content and feature updates to support these goals should be scheduled out and planned in advance. Of course, if you receive any bug reports or technical issues from your users, integrate those fixes into your release schedule. The production team should also be conscious of something called tech debt, which has a tendency to creep up within projects that are continuously updated over time. Tech debt is accrued over time as updates are made to your product, distancing development from the original libraries and software that was used. As you work on outlining your road map and setting up the development work in your task-tracking platform, tag tasks related to content updates, feature updates, bugs and issues, and tech debt accordingly.

Content Updates

Scheduling content updates can extend the life of your application and retain your audience. These updates come in the form of introducing more difficulty or complexity to the original content or by adding new modules or packs to extend the XR experience into new subject matter. If you plan on implementing more challenging content but with the same subject matter, build complexity upon what you've already created. Reengage with your subject matter experts that helped you design the original content and identify additional details or difficulty associated with the steps or processes in the XR experience. This will add dimension to your existing content so that your users can return and experience something new, while staying true to the original intent of your program.

If it's a better fit to evolve the next phase of your XR experience by adding completely new modules, information, or storylines, start building out your design documentation to support this. As you add new content, recall the formatting and standards you set for your original designs. As you're adding on to an existing XR experience, you want to align with the style and composition of your original application so that the updates make sense to the user. This will lead to a cohesive experience that could support infinite expansion over time.

Feature Updates

You're bound to have users make feature requests after deployment. Once people experience an extended reality application, their imaginations run wild, and they often verbalize their ideas in the form of feedback. This is a great sign as your audience is engaged and wants more. In one of our public launches, we published a feature-tracking board where our audience could see the status of their feature requests and which ones were going to make it into a future update of the XR experience. They also had the opportunity to upvote other people's features, which helped us prioritize which elements would be most meaningful to our audience. That level of transparency isn't a requirement, but it can be helpful to your own production process when prioritizing.

As you incorporate feature updates, consider which of these interactive elements or quality-of-life improvements are going to make the biggest impact on the user experience. Identify which tasks will help your audience continue to come back to your extended reality deployment time and time again. Sometimes adding new interactivity isn't as powerful as quality-of-life updates, which make for a more comfortable and easy-to-use application. However, adding new interactive elements can create more excitement and buzz-worthy discussion about your XR experience, which may also have a positive impact on the longevity of your product. Regardless of which features you choose to update, keep your audience and their anticipated reaction of these elements in mind when you prioritize them in your release schedule.

Tech Debt

While technical debt, or more commonly referred to as tech debt, is not a fun subject, it's integral to the long-term success of your extended reality application. Any digitally produced project is going to be reliant on the systems, programs, and libraries it used in its original production. Unfortunately, with the ever-changing landscape of extended reality tools and programs, these systems can become outdated or deprecated as time goes on. Take note of the systems your project actively uses, and pay attention to any changes or discontinuations that affect these systems as you make updates to your XR experience.

At one point, it will become important to address the tech debt your project is taking on, or the development files may become unstable or unusable. This has happened to our team many times and will likely happen to yours as well if the product stays active long enough. For example, we produced a mobile augmented reality application for iOS and Android consumer devices. For iOS, we leveraged the tools ARKit and SceneKit, which are still around and thriving to this day. For Android, we used their comparable solutions, ARCore and SceneForm. Years into the project, SceneForm was set to be deprecated. We had to find a new solution after the app was already on the market to replace SceneForm, which served as the 3D renderer for the entire Android deployment of the application. This transition took months and was not on our original road map but allowed us to eliminate tech debt and continue making updates moving forward.

Whether your project is deployed for only a week or will see continuous updates for years to come, you are producing something within an emerging technology field, which is no small feat. The power of extended reality and the opportunities that exist are vast, leaving much uncharted and ripe for innovative thinkers to make their mark. We're still in the Wild West of an ever-expanding medium, and we've only scratched the surface of what's possible. One of the biggest drivers of adoption and forward progress in the field of extended reality is content development. I do not want to be a gatekeeper, because that would not help the industry advance in the way that it needs to succeed. So take this extended reality blueprint and make it your own. I cannot wait to see what you create.

16

The Future

IF THERE IS one thing I've learned by being a small part in this incredible world of extended reality, it's that you must always keep your eye on the future and what's coming next. Information and technology advancements are accelerating so quickly that it's difficult to keep up with the latest and greatest. There are some channels I use to keep an eye on what's coming next, whether it's newsletters, conferences, events, or speaking with my professional peers. However, some of the best practices that help me keep an eye on the future are found by reaching into the past and learning from it.

Entering XR Professionally

Getting into the extended reality field today is both more difficult and easier than it was a decade ago when I started. At the time I was introduced to the technology, I could just push my way in and figure things out as I went. There wasn't a great deal of competition, and while there were skeptics in my network, overall, the sentiment was welcoming. Now, there are more tangible opportunities, but these opportunities also require more experience. There are some parallels that can launch easily into a career in extended reality, such as gaming, user experience design, and creative agency backgrounds. Almost any industry can leverage extended reality, and as you've learned throughout this book, jobs are not limited to developers and digital artists. There is a need for designers, project managers, audio engineers, and any role that exists within other organizations such as human resources, opera·tions, finance, and administrative personnel.

If you're truly interested in entering the field as a profession, create your own independent projects, construct a well-rounded portfolio, and gain experience by engaging with XR enthusiasts and professionals in your area. Don't wait around for the perfect opportunity—create it yourself. If you're interested in entering the field in a nontechnical role, still engage with the community around you, do industry research, write blogs, and experience everything you can get your hands on. There is a place for you here if you're interested; you just may have to get creative to find it.

Speed

Everything is happening so fast it's hard to keep up with all that's being deployed on the market, both on the hardware and software side of the industry. When it comes to hardware, it can be difficult to know which devices to invest in. We've purchased or received every headset you can imagine over the years, and some of them unfortunately sit on a shelf collecting dust. When my developers ask if I want to sell them, I joke, "Those are for my museum!" Looking back on the headsets that haven't worked, they seem to suffer from one of two issues. First, they were created for only one purpose or one experience. No one wants to buy a hardware device for a single application—it's unrealistic. In fact, I really hope that trend is over. Second, they had design flaws that resulted in a poor user experience. Most of these headsets were cranked out quickly to make a buck and attempt to compete with the biggest hardware manufacturers in the world. They failed because of it, so while speed is important, it doesn't replace a good product.

On the software and content side of speed, I'd actually argue the opposite of my feelings about hardware—do things faster, and don't take too long to make a decision. While a polished product is going to require massive attention to detail and lots of time, prototyping interactions and creating proof-of-concept-level experiences should happen fast. Even if you fail, you learn something. Failing fast is more beneficial to overall progress than spending too long creating something perfect and missing your window of opportunity. The industry moves quickly, and so do the opportunities and excitement associated with it.

Patterns

I've found throughout the years that extended reality technology is extremely cyclical. In 2015, 100 percent of the work Futurus produced was virtual reality or 360-degree video content. In 2017, my company's work consisted of 80 percent augmented reality production. In 2022, the company's project load was inversely about 75 percent virtual reality work. I've seen these trends flip-flop over the years, and I'm hopeful that instead of another flip, virtual and augmented reality applications will start to merge. More virtual reality devices are starting to introduce passthrough and the ability to still use the 3D graphics while in passthrough mode. Some newer augmented reality or spatial computing headsets are promising the ability to dim or occlude the lenses entirely, allowing you to see virtual overlays in near full opacity. The merge is coming, so keep an eye on opportunities where crossover content may be a fit, and start to plan for it.

I've also observed patterns in what type of content people are requesting. As a company that other organizations hire to produce their extended reality applications, I get to hear it all. From 2015 to 2018, 360-degree video tours were extremely popular. This also happened right around the time Google Cardboard was released and sent out to hundreds of thousands of households by the *New York Times*. I would argue that this campaign was the most powerful driver of virtual reality adoption in history. While it wasn't "true VR," nearly overnight it educated millions of people about virtual reality and what it could become. The only downside was that 360-degree video presented to users with a phone and a cardboard box made quite a few people feel ill. This turned off countless people from exploring the technology further, and only in the past few years with the emergence of mobile 6DOF devices are those individuals starting to come back around to trusting the technology.

Similarly, I mentioned earlier that 2017 was our biggest year for augmented reality projects. Well, what happened in July 2016? Pokémon Go was released to the public and took augmented reality to great success and name recognition. After the wildfire of that game started to pick up, that was around the same time large corporations were planning their 2017 budgets, and what do you think they considered seriously for the first time? Augmented reality. We also no longer had to spend the first part of every meeting doing an educational overview of what the technology was. People just knew, and that was a relief. Keep an eye on the patterns and trends you see in the market, as they may make a big impact on your extended reality production cycle.

Observe

Learning from others and observing how people in the extended reality and adjacent industries are creating and innovating is some of the best education you will ever receive. Speaking to your peers and facilitating intelligent conversation about industry trends, findings, and experiences will spark ideas that you would not have otherwise thought about. I've been able to foster these conversations by staying active in my local community, as well as networking with people all around the world in the industry. Start building your network as early as possible. I find that most people want to connect and engage with each other and share what they're working on. Those of us who are deep into extended reality understand that it's going to take collaboration for the industry to take off in a way that becomes ubiquitous in society.

We try to experience anything we can get our hands on. This not only means hardware but also software and experiences that others produce. You can learn tons about what works and what doesn't by observing how others have implemented interactivity and functionality within their own projects. As you go through other people's projects, note what interactions feel good and indicate which features you'd like to consider incorporating into your own project. The more consistency we can get with interactions across projects, the easier it will be for newcomers to adopt the technology. Tutorials are important, but as mentioned in the design chapter (Chapter 5), we can't just make a tutorial for our experience: we must also educate new users how to use the technology. When interactions and expectations in design become consistent, we will no longer have to educate on how to use the technology and will be able to focus solely on teaching our audience how to use our XR experience.

Something that my team has scheduled on a recurring basis is called "Game of the Month." The team gets together on the first Friday afternoon of every month and enjoys an extended reality game or experience. Most people play firsthand, but we also stream to the entire team in case the game has limiting factors (such as smooth locomotion, which can make some of us sick). While playing these games, we observe how other studios are incorporating immersive elements and take note of which may be a good fit for future projects. We also document when something isn't working so that we know what to avoid in our own production process. The following Monday, we all share our findings in a roundtable discussion so that each person's thoughts are heard. While my organization isn't making games per se, there is so much of what we do at the root of gaming. The gaming industry also sees most of the good stuff first, from the latest hardware to graphics advancements to new development tools. If we're not learning from the gaming industry, we will fall behind, which is detrimental in an emerging technology field. Some of our clients don't like to think of what we do as gaming, but we find the best forecast of what's to come is inspired by this powerful industry.

Adjacent Technologies

We can learn a great deal regarding what's to come by looking back at our past and engaging with members of the extended reality community. However, staying on top of the latest technology also requires a keen eye on the active pulse of the industry. There are also adjacent technologies that are starting to merge with extended reality production, such as artificial intelligence or machine learning. It can seem overwhelming to stay on top of it, which is why I like to subscribe to industry aggregate newsletters. By looking at the top extended reality headlines for the week, I can get a snapshot of what projects are launching, hardware release dates, investments in new technology, and innovative breakthroughs. Diving deeper into relevant articles or press gives me a greater understanding of what's going on now and can help influence some of my business or production decisions. Keeping an eye on adjacent industries can be helpful when piecing together the big picture of the future of immersive technology.

Artificial Intelligence

Artificial intelligence, or AI, is everywhere. It's in every tech conversation, and companies are already actively using the tools available to them. I can't even tell you how many people asked me if I used AI to write this book, which I firmly and intentionally did not. So how does that overlap with extended reality production? It can have some impact on the production process, but proceed with caution as integrations and implementations are still nascent. We have started to use AI voice services to generate temporary voice-overs while we are working through storyboarding and initial implementation of narrative elements. Not all AI voices have realistic human emotion just yet. So unless it's very straightforward instruction, we want to use human actors to get that emotion across. The experience is more immersive when the characters have the feeling of a true human.

My developers will also sometimes use AI systems, such as ChatGPT, to inspire them to think of a solution in a new way. We have a rule that no one can use a solution delivered from an AI system without verifying it and running it by other members of the production team. These systems leverage outdated and sometimes incorrect data to provide their answers. Hopefully this will not be the case long-term, but we need to still leverage the knowledge that my team has acquired over time to confirm if a suggested solution will work. The benefit of leveraging these systems is that they often provide a new perspective or suggestion that the developer just couldn't see before. It's a good thing overall, when paired with the oversight of an experienced professional.

Moving into the future, I'm most excited about the evolution of AI-driven tools that will help optimize the art and environmental assets that go into an extended reality experience. There are already tools on the market that will create 3D assets for you, including textures, which will continue to thrive as well. The biggest bottleneck in extended reality adoption is content production. It takes too long to produce content to keep up with the appetite of the masses. If we're able to expedite the pipeline of extended reality production with tools for creators, it will be easier to make content, and adoption will surge.

Machine Learning

While machine learning, or ML, can be considered a subset of artificial intelligence, it deserves its own callout when speaking about the future of extended reality technology. I find machine learning to be exceptionally important to the future of augmented reality and spatial computing due to the advancements of computer vision. Computer vision is the ability for our digital devices and their cameras to identify and understand the things around us. This can include text, images, objects, and even people. When directly applied to an augmented reality wearable, the computer vision system can identify things around you and deliver related content as a digital overlay onto the physical world. For most augmented reality experiences today, physical objects and images must be programmed to respond directly with what content is overlaid. As computer vision improves, this will require less up-front programming and more reactive content in our direct line of sight.

Another subset of machine learning is natural language processing, which indicates the ability for a computer system to intake speech of a human, interpret it, and deliver a response. The machine learning piece of this is that the system will be able to make predictions based on patterns, and it will continue to learn as the system receives more information. Ideally the system will become smarter, but that all depends on the information it's able to take in. Natural language processing can lead to more realistic interactions with NPCs in an extended reality experience. It can also impact dynamic scenarios in an experience that provides situational content. Once again, these tools have the potential to affect the speed at which content can be produced, which will drive adoption.

Metaverse

The term "metaverse" has been floating around since the 1992's science fiction novel *Snow Crash*, yet it's picked up traction in recent years, for better or worse. The biggest grab at the word occurred when social media giant Facebook changed its entire company brand to Meta in 2021, going all in on nabbing the name recognition for itself. Around that same time, every company with a flexible innovation budget tried their hand at entering the metaverse, and it became an inauthentic exercise in technology hype. The metaverse situation negatively affected the extended reality industry by overinflating expectations, which many of these brands could not deliver on. I realize this sounds pessimistic, but I had a front row seat watching it all go down and saw how it directly affected my business and the expectations of the technology I had dedicated my career to.

Now that the hype has died down, I would like to go on record that the metaverse is not dead as many have claimed. The individuals who were working on building the future of the Internet and its next phase are still around. They started on this long before Facebook changed to Meta, and they are still working on creating these shared spaces where individuals can connect, work, relax, and build online together. The metaverse has many different definitions, but I see it as the connection of shared digital spaces where user-generated content is at the root of progress. The ideal metaverse will support the interoperability of objects and content across spaces. This would mean that an object you have in one space can be carried with you to another independent space.

Interoperability as has been promised is still a long way off due to technical limitations and proprietary standards many of these separate spaces have. However, just like you can drive a car across state lines, someday you will be able to bring your collection of personal digital objects across the Internet.

Hardware Forecasting

Keeping an eye on the future of extended reality hardware isn't as difficult as it sounds. Most hardware manufacturers announce a device is in production well before the release date, which gives us visibility into what's coming up. This goes for headsets as well as haptic devices. And another benefit of these announcements is that they provide developer tools and SDKs much earlier than the physical device is available. When thinking about the future of your projects, consider the production timeline and what will be available at the end of the project. If you're developing something for one brand of headset and the latest version of that headset will be available by the time the project is complete, will you deploy it on the newest version of that device? It's a question worth considering, especially as the hardware cycle in XR is about two years of life for most of these headsets.

The challenge with planning a project for release with a specific headset that hasn't hit the market is the very real possibility of delays. Whether it's due to supply chain issues or the operating system itself still being worked out, hardware delays are common with extended reality devices. Plan as much as you can, but understand that certain things are going to be outside of your control when it comes to hardware. We have worked on integrating the SDK for a pre-released device that ended up being scrapped altogether. Surprising and unfortunate, but we were able to pivot and release on a comparable device that was available on the market at the time. Some of the features that new devices promise may also be rolled out over time. For example, as devices with enhanced passthrough capabilities have been released, the developer tools to access the passthrough function in a project weren't made available until months later. In cases like this, the manufacturer still implements the hardware required to support these features but spends additional time finalizing the functionality and developer tools after the release. New devices are coming to the market all the time, and competition is good. So keep your eye on these announcements, and hold your breath until the devices are actually in hand.

If you've learned anything from this book, I hope it's that there is much to create and accomplish in the field of extended reality. There is room for anyone with an interest and the drive to produce something unique. While my journey has been met with bouts of frustration and puzzled looks from friends and family who didn't get it at first, I could not be happier to be in the position I find myself in. I get to watch extended reality evolve into the powerful medium that it is daily. So, whether you picked up this book because you want to enter the industry yourself, or you just wanted a better understanding of the technology and how these immersive experiences come to fruition, I'm glad you found my extended reality blueprint. Through education and understanding comes adoption and advancement, which is what we need to move this industry forward. To the future!

Index